SCHOOL OF GEOGRAPHY
A332

AF616314

EF61W

Springer Series
in Physical Environment

8

Volumes already published

Vol. 1: Earth Surface Systems
R. Huggett

Vol. 2: Karst Hydrology
O. Bonacci

Vol. 3: Fluvial Processes in Dryland Rivers
W. L. Graf

Vol. 4: Processes in Karst Systems.
Physics, Chemistry and Geology
W. Dreybrodt

Vol. 5: Loess in China
T. Liu

Vol. 7: River Morphology
J. Mangelsdorf, K. Scheurmann and F. H. Weiß

Vol. 8: Ice Composition and Glacier Dynamics
R. A. Souchez and R. D. Lorrain

Volumes in preparation

Vol. 6: System-Theoretical Modelling in Surface Water Hydrology
A. Lattermann

Vol. 9: Desertification.
Natural Background and Human Mismanagement
M. Mainguet

Vol. 10: Fertility of Soils.
A Future for Farming in the West African Savannah
C. Pieri

R.A. Souchez R.D. Lorrain

Ice Composition and Glacier Dynamics

With 119 Figures

Springer-Verlag
Berlin Heidelberg New York
London Paris Tokyo
Hong Kong Barcelona
Budapest

Dr. ROLAND A. SOUCHEZ
Dr. REGINALD D. LORRAIN

Université Libre de Bruxelles
Faculté des Sciences, C.P. 160
Département des Sciences
de la Terre et de l'Environnement
Avenue F.D. Roosevelt, 50
1050 Bruxelles, Belgium

Cover illustration: The margin of Sydkap Ice Cap (S-W Ellesmere Island, Arctic Canada). The upward movement of ice is visible when not masked by a snow cover. Debris is transferred from the ice-bedrock interface towards the ice cap surface and forms an ice-perched moraine. The composition of these basal ice layers gives some clues about their formation. (Photo by R. Souchez)

ISSN 0937-3047
ISBN 3-540-52521-1 Springer-Verlag Berlin Heidelberg New York
ISBN 0-387-52521-1 Springer-Verlag New York Berlin Heidelberg

Typesetting: K+V Fotosatz GmbH, Beerfelden
32/3145-543210 – Printed on acid-free paper

To E. Picciotto,
who introduced us into the field
of isotope glaciology

Preface

Ice composition has until now been mostly used for reconstructing the environment of the past. A great research effort is made today to model the climate system in which the ice cover at the earth surface plays a prominent role. To obtain a correct model of the ice sheets, due attention must be paid to the physical processes operating at the interfaces, i.e. the boundary conditions. The idea behind the title of this book is that the study of ice composition can shed some light on the various processes operating at the ice-bedrock and ice-ocean interfaces and more generally on glacier dynamics. The book is not intended as a treatise on some specialized topic of glaciology. It is mainly the product of the experience of the two authors gained over several years research on the subject.

The two authors are both members of the same university department and personal friends. The book was prepared in the following way. After a first draft of the complete book had been written by the first author, it was put in the hands of the second. The final version sent to the publishers is therefore the result of extended discussion, while at the same time preserving the unity of style that would have been lost had the two authors written selected chapters of the book individually.

The book is organized into two distinct parts. The first part is devoted to fundamentals which must be understood before attempts at ice composition study can be undertaken from the perspective of glacier dynamics. The reader is first introduced to the glacier system, its main characteristics and how it works (Chap. 1). The various ice types which are produced within the glacier system by specific processes of formation are considered. Chapter 2 is then devoted to ice composition with emphasis on isotopes in ice. Phase equilibria, self-purification and leaching are considered in order to understand the distribution of impurities in ice. Ice also contains mineral particles and gas bubbles, representing a tiny part of the atmospheric reservoir imprisoned during ice formation as the pores of the firn close off and entrap the air. These factors are considered as they can also help to understand glacier dynamics.

The second part of the book is concerned with the implications on glacier dynamics. A general view of the relationships between ice composition and ice flow is first considered in Chapter 3. A more detailed view of the key basal zone of ice caps, ice sheets and alpine glaciers is then given in Chapters 4 and 5. In the polar regions, glacier ice comes into contact with the sea (Chap. 6). This contact zone between glacier and ocean plays an important environmental role. It is considered here since ice composition studies further the understanding of floating ice dynamics. Finally, implications at the global scale are considered.

The authors are aware that important work may have been omitted and that imperfections may remain in the text. Their only excuse is that they can devote only a limited amount of time to this activity, as they are continuously involved with ongoing research. Whilst apologizing for any inexactitudes, the authors sincerely hope that this book will serve the cause of glaciology.

This book would probably never have been written without the support of many individuals. Professor D. Barsch from the University of Heidelberg is gratefully acknowledged for encouraging the authors to write this book. The help of Dr. M. Lemmens, who reviewed a first version of the manuscript, is also appreciated. The work greatly benefited from considerable improvement of the English and constructive comments by Dr. M. Sharp and Mr. B. Hubbard from the University of Cambridge during a stay at the authors' laboratory. Critical comments of Dr. D. Wagenbach from the University of Heidelberg resulted in significant improvements of the structure of the book. The tables and the drafts of diagrams for this book were prepared by Mrs. J. Escande. Most of the typing was done by Mrs. E. Rondou. Many thanks are due to them both for their skill and patience.

Brussels, Winter 1990/91

R. A. SOUCHEZ
R. D. LORRAIN

Contents

Part I Fundamentals

1 The Glacier System 3

1.1 General 3
1.2 Input and Output 4
1.3 Ice Flow 8
1.4 Ice Residence Time 13
1.5 Distribution of Ice Fabrics and Textures 14
1.6 Ice Foliation 18
1.7 Basic Ice Types in Different Glacial Systems 22

2 Ice Composition 29

2.1 Stable Isotopes and the Water Cycle 29
2.2 Stable Isotopes in Snow 33
2.3 Isotopic Changes During the Transformation of Snow into Ice 39
2.4 Stable Isotope Fractionation by Freezing 45
2.5 Impurities and Phase Equilibria 60
2.6 Self-Purification and Leaching of Impurities 64
2.7 Mineral Particles in Ice 68
2.8 Gases in Ice 74

Part II Implications on Glacier Dynamics

3 Ice Composition and Ice Flow: a General View 79

3.1 Gas Content and Ice Sheet Profiles 79
3.2 Impurities in Ice and Ice Creep 83
3.3 Isotopes and Flow in Ice Sheets and Ice Caps 86
3.4 Lead 210 in Ice and Alpine Glacier Flow 97
3.5 Mineral Particles in Ice and Glacier Flow 101
3.6 Evidence for Buried Glacier Ice 108

4 The Basal Zone of Ice Caps and Ice Sheets 114

4.1 Thermal Conditions at the Glacier Sole 114
4.2 The Effective Bed 117

4.3 The Basal Zone in Ice Cores 123
4.4 Investigations in Marginal Areas 130

5 The Basal Zone of Alpine Glaciers 140

5.1 Water Flow in the Basal Zone 140
5.2 Phase Changes at the Base of Alpine Glaciers 143
5.3 Incorporation of Debris into Basal Ice 145
5.4 Basal Ice Chemistry 148
5.5 Subglacial Precipitates and Basal Ice 155
5.6 Isotopes in the Basal Zone of Alpine Glaciers 158

6 The Contact Zone Between Glacier and Ocean 164

6.1 Ice Shelves and Tidewater Glaciers 164
6.2 Melting and Freezing at the Base of Ice Shelves 166
6.3 Frazil and Congelation Ice 170
6.4 Isotope and Impurity Distribution 172
6.5 The Case of the Ward Hunt Ice Shelf 177
6.6 Freezing Rates in the Marine Environment 179
6.7 The Glacial Supply to the Ocean 183

Conclusion: Ice Composition, Glacier Dynamics and Global Changes 187

References 190

Subject Index 201

Part I Fundamentals

1 The Glacier System

1.1 General

A glacier, whether it be a valley glacier, an ice cap or an ice sheet, is an ice mass at the surface of the earth. It can be considered as an open system with input, storage, transfer and output of mass. It is a system in dynamic equilibrium where the mass balance depends on input and output. Input or accumulation includes all those ways in which mass is added to a glacier: solid precipitation, wind-drift snow and avalanching, growth of superimposed ice, freezing of water and condensation of vapour to the ice surface. Output or ablation includes all the ways in which mass is lost from a glacier: melting, evaporation, wind deflation and iceberg calving being the most important. The balance is the difference between accumulation and ablation over the entire glacier for one year. The area in which there is an excess of accumulation over ablation, or net accumulation, is called the accumulation zone, while the area in which there is an excess of ablation over accumulation, or net ablation, is called the ablation zone. The boundary between the two zones is given by the equilibrium line where, over a year, ablation equals accumulation.

Glacier ice is formed in the accumulation zone mainly by transformation of snow into ice. However, it exists below the equilibrium line, since glacier ice is deforming and flows under the influence of gravity. The extent of a glacier below the equilibrium line is dependent upon net ablation and ice discharge. If, at the glacier front, the supply of ice per year from upglacier is equal to the net ablation and such a balance is maintained over a number of years, the front will be stationary. If ice discharge is larger, the front will migrate downslope and this will give rise to an extending glacier. If, on the other hand, less ice flows from upglacier than is destroyed by ablation in the frontal zone, then the glacier will retreat. However, in all cases, ice always continues to flow downslope. A change in the balance can thus induce extending or receding conditions.

Ice sheets and ice caps develop when snow accumulation, transformed into ice, exceeds the capacity for drainage by valleys. Topographic irregularities are thereby progressively buried. The glacier reservoir becomes more important as the altitude of the equilibrium line falls during the process. The white surface of snow has a higher albedo than the pre-existing land surface. Thus, more solar energy is reflected and this slows down ablation rates. When the ice mass is large enough, the atmosphere in contact with the ice is cooled to such a degree that a reduction of solid precipitation occurs. A low precipitation input is characteristic of ice sheets. However, as ablation is also low, equilibrium lines lie in their peripheral zones.

1.2 Input and Output

The first step in the formation of glacier ice is snow fall at the surface of the glacier. Transformation of snow into ice is a metamorphic process that occurs at low pressure and low temperature. This transformation can occur in the presence or absence of meltwater. During metamorphism, the density is increased so that the ice stage is reached at a density of about 0.85 g/cm^3, when the air is present only in the form of gas bubbles and no longer in pores more or less connected to the outside atmosphere. The intermediate stage between snow and ice is called firn. It is reached when the density cannot increase any more by settling alone.

Transformation of snow into ice is a slow process in the absence of meltwater. From a knowledge of the accumulation rate (in water equivalent) and the depth of the firn-ice transition, it is possible to compute the time taken for ice to form. Values given in Paterson (1981) for Antarctic and Greenland stations range between 80 and 4000 years. Under colder conditions, the firn-ice transition is deeper and the transformation takes longer. An extreme value of 4000 years is reached at Vostok in East Antarctica but more "normal" values for polar areas are between 100 and 200 years. By contrast, as cited by Paterson (1981), transformation of snow into ice on Seward glacier in Alaska where meltwater is present, takes only 3 to 5 years, and the firn-ice transition occurs within 13 m of the surface, as compared to 70 m at Byrd Station in Antarctica.

In the absence of meltwater, sublimation or direct transformation of water vapor into ice plays the major role in the ice formation process. The second law of thermodynamics implies that the free energy of a system tends towards a minimum value. Reduction of total crystal area will reduce free energy. So, by sublimation, grains acquire a more globular shape and big crystals will have a tendency to grow at the expense of small ones. Settling also plays an important role in this early stage. Globular particles settle more readily up to a density of about 0.55 g/cm^3. Packing experiments indicate a lower limit for porosity to be about 40%. Such a porosity for spheres of ice of density 0.9 g/cm^3 corresponds to an overall density of 0.55 g/cm^3 (Paterson 1981). Sintering, or the transfer of matter to the points of contact between ice crystals, forms bonds and further reduces the total surface area. With the increase in density, the firn is less porous and sublimation greatly reduced. Molecular diffusion now takes the dominant role in reducing stresses on the grains by changes of grain size and shape. Recrystallization occurs, which can increase the density up to that of ice, by progressively closing the pores of the firn until all links with the outside air are sealed. A further increase in density is still possible by compression of gas bubbles in the ice, the pressure within which may reach a few atmospheres.

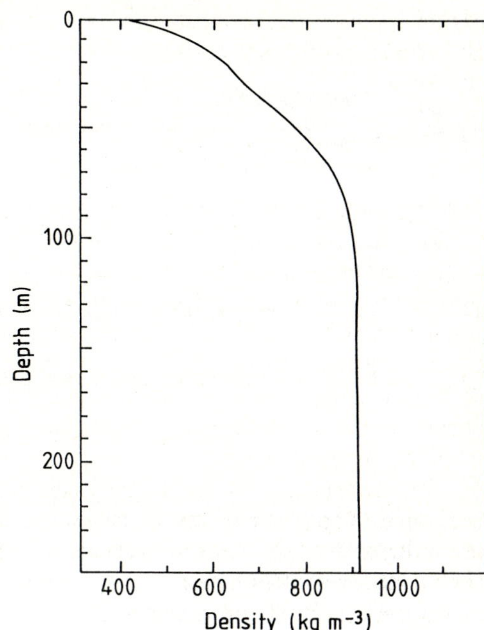

Fig. 1.1. The increase in firn/ice density with depth at Byrd Station, Antarctica. (After Fig. IV-2 in Mellor 1964)

Figure 1.1 gives the increase in density with depth at Byrd Station in Antarctica. The change in gradient of the curve at about 20 m depth is due to the formation of firn, corresponding to the end of the settling process. Ice is reached at a depth of about 70 m when the density is 0.9 g/cm^3.

In some circumstances, a coarse grained type of firn called hoarfrost may be formed in the absence of meltwater. Crystals in the range of 0.2 to 0.5 cm are produced within the snowpack by sublimation. A strong thermal gradient producing a strong gradient of vapour pressure is needed for the development of this hoarfrost.

When the melting point is reached, grains readily become rounded, since they melt first at their extremities and smaller grains tend to melt before the larger ones. Meltwater accelerates packing by lubricating the grains. Refreezing of meltwater at depth in the snowpack fills the air spaces and so a speed-up of the general process takes place: snow is transformed into ice in a few years.

If one considers a layer of one year's snow accumulation at the glacier surface, different zones can be distinguished, depending on the amount of meltwater produced at the surface. These different zones are not necessarily present on a single glacier but large ice masses usually develop more than one zone. In the *dry-snow zone*, the transformation of snow into ice occurs below the melting temperature, in the complete absence of meltwater. This is due to very low air temperatures. Such a situation exists today in the central parts of the main ice sheets, either in Antarctica or in Greenland. In the *percolation zone*, some surface melting occurs, but the water produced refreezes at depth in the cold snow layer of the previous winter. Latent heat is released during re-

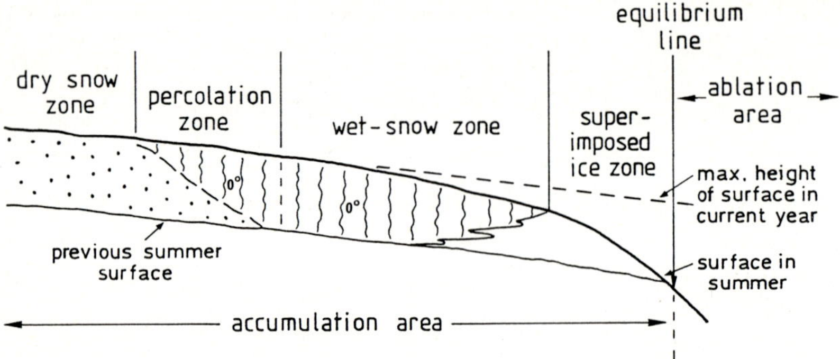

Fig. 1.2. Zones in the accumulation area of a hypothetical ice sheet or ice cap. (After Fig. 3 in Müller 1962, p. 305)

freezing and this warms up the snow. Freezing of 1 g of water releases enough latent heat to raise the temperature of 160 g of snow by 1 °C. When more meltwater is produced, as in the *wet-snow zone*, the entire snow cover deposited since the end of the previous summer is raised to 0 °C within one year of deposition. Superimposed ice is formed when the meltwater comes into contact with the previous year's cold surface at depth and refreezes. In the *superimposed ice zone*, by the end of the summer, the snow layer is partially melted and transformed into a superimposed ice layer. This is a special case where transformation of snow into ice takes less than one year. Figure 1.2 gives a profile of the different zones on a hypothetical ice cap.

Cold ice forms when surface melting in the summer is negligible or absent. In such a case, the temperature at a depth of 10 m, i.e. below the level of seasonal temperature variations, closely approximates the mean annual air temperature at the site. In such a cold ice mass there is a general increase in ice temperature with depth due to geothermal heat. The rate of this temperature increase is strongly influenced by the accumulation rate. Figure 1.3 from Robin (1955) gives the relationship between ice temperature and depth for a hypothetical 3000-m-thick ice sheet under various values of snow accumulation in water equivalent.

A second situation leading to cold ice formation is linked to the cooling of the upper part of a glacier by winter cold. If a greater thickness of ice is cooled in winter than can be warmed in summer, then a cold layer near the surface may survive the following summer.

Temperate ice, that is ice at the pressure melting point, is formed when there is sufficient meltwater produced, to raise the annual accumulation layer to 0 °C. On the other hand, basal heat sources may be sufficient to raise the temperature of the glacier base to the pressure melting point. Geothermal heat and heat released by friction are the principal sources of basal heat. In such a situation, water is likely to be present at the bed because a given amount of geothermal heat or frictional heat can be used to melt a thin layer of basal ice. When

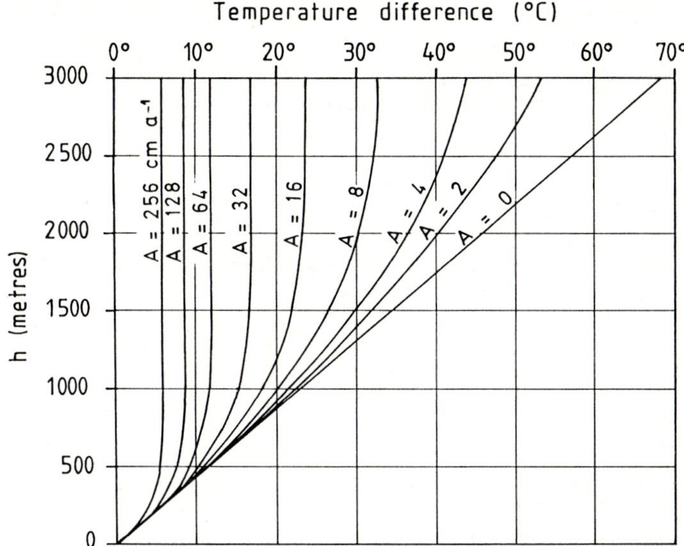

Fig. 1.3. Temperature difference between the base and other levels near the centre of a hypothetical ice sheet with a central thickness of 3000 m. *A* is the accumulation rate (Robin 1955, Fig. 2)

a basal layer of temperate ice exists, which means no temperature gradient in the basal zone, then all the geothermal heat and the heat released by friction are used in basal melting and no heat is conducted away into the ice.

Independently of snow metamorphism, the growth of superimposed ice and the freezing of water, the condensation of vapour directly onto a glacier surface may also represent a significant input under some circumstances. Rime ice can be formed in cool maritime glacial environments; cauliflower-shaped masses of rime build up in the Antarctic Peninsula on exposed nunataks and fall to the glacier below. Koerner (1961) calculated that rime ice contributed 2.5% to the total accumulation in the region of the Antarctic Peninsula he has studied. However, this contribution to glaciers or ice sheets will not be expanded upon in this book.

Output is the mass loss from the glacier system considered. Melting, sublimation, wind deflation and calving are the principal output processes. However, if surface melting is followed by percolation in the snow and then by complete refreezing within the basal part of the snow cover, a process which commonly occurs in the percolation zone, this is not considered as output, since there is no mass loss for the glacier. Most ablation occurs by surface melting. Although some ablation is achieved by sublimation – in this case, the direct phase change from ice to vapour – this process is generally of little importance except when total amounts of ablation are small as in, for example, a cold Antarctic continental climate.

Melting in a glacier occurs not only at the surface but also in deep ice layers as a consequence of heat released by friction during ice movement and at the glacier base as a result of the geothermal heat flux. Water produced by such internal and basal melting can be evacuated from the glacier. The water can be transported by the subglacial drainage system to the frontal zone where it leaves the glacier system. Another possibility is for the water to percolate through the subglacial sediments or through permeable subglacial bedrock into an aquifer.

On any glacier, there is a tendency for net ablation (ablation rate minus accumulation rate) to be the highest in the frontal zone. A stationary front means a balance between net ablation and ice discharge. Ice flow is thus an important component of the glacier system.

1.3 Ice Flow

Ice is not characterized by a distinctive yield point but rather deforms gradually. Careful laboratory studies of the deformation properties of polycrystalline ice made by Glen (1955) have shown that strain rates vary as the third power of the applied stress as well as being dependent on temperature. In its simplest terms the flow of ice reduces to the following basic power relation:

$$\varepsilon = D \cdot \tau^{n} \ ,$$

where ε is the effective strain rate, τ is the effective stress, n is a constant (= 3) and D is a thermally activated ice hardness factor. The thermally activated ice hardness factor is given by:

$$D = D_0 \, e^{-F/RT} \ ,$$

where D_0 is a constant independent of temperature, R is the gas constant, T is the absolute temperature and F is the ice activation energy. This equation suggests that under conditions of constant effective stress, the logarithm of the strain rate is directly dependent on the temperature. Above about $-10\,°C$, polycrystalline ice softens appreciably. The creep rate being unaffected by cryostatic pressures, this provides the physical basis for rejecting old theories of extrusion flow in glaciers.

The large-scale flow of ice sheets and glaciers is governed by this non-linear flow law. In addition, where conditions at the bed are favourable, basal sliding may occur and may account for a significant proportion of the motion. The question of how much of the motion of ice sheets is due to internal deformation and how much to sliding has not been fully solved, although a certain number of indications exist.

Nye (1951, 1952) has shown that the thickness of an ice sheet at any point is inversely proportional to surface slope and the basal shear stress τ_b is given by:

$$\tau_b = \varrho g h \sin\alpha \ ,$$

where ϱ is the density of ice, g is the acceleration due to gravity, h is the ice thickness and α is the angle of slope. Field investigations on ice sheets indicate that this equation only applies to mean slopes calculated over distances an order of magnitude greater than the ice thickness. At shorter distances, the surface slope of ice sheets is strongly influenced by gradients of longitudinal stress.

This standard equation for basal shear stress implies that the bedrock friction exactly balances the downslope component of the weight of the overlying ice. This is not the case everywhere. There are regions where the downslope component of the weight exceeds the friction; in such a region the ice pushes the downstream ice and pulls the ice from upstream. Similarly, in regions where the friction is greater than the downslope component of the weight, the ice mass is pushed by the upstream ice and pulled by the ice below. In such regions, as indicated by Paterson (1981), the longitudinal stress component is important and a correction term has to be added to the equation for the basal shear stress.

τ_b, when calculated over a large area is also considered by some authors as a driving stress, since the equation fundamentally describes the force required to induce motion in the ice. Comparing values of this driving stress with the ratio of velocity to thickness, it can be seen that the motion of the central part of an ice sheet frozen to its bed is largely due to internal deformation. The ratio of velocity to thickness is proportional to the mean shear strain rate through the ice column in the absence of sliding. Budd and Smith (1981) established a regression line between the driving stress and this ratio for polar ice which is thought to be moving largely by internal deformation. Data from McIntyre (1985) plotted on Fig. 1.4 for the upper part (upstream of a step) of the Byrd Glacier profile in East Antarctica indicate that creep is the main process of flow in that area, although the scatter of points is relatively high.

The basis of such a steady-state model of flow is the following. Given a mass input of ice to the surface, an ice sheet will develop surface slopes which create the stress required to allow ice to flow outwards and thus balance accumulation. This is a form of the continuity equation expressed by Drewry (1986) as

$$\frac{dq}{dx} + \frac{dh}{dt} = A(x, t) ,$$

where q is the mass flux or discharge in the x or downglacier direction, h is the ice thickness, t is the time and A is the accumulation.

Now, if the width of a glacier is not much larger than its thickness, then the effects of valley sides must be taken into account. Raymond (1971) shows how surface velocity varies across a cross-section of Athabaska Glacier, Canadian Rockies (Fig. 1.5). The velocity changes little across the central part of the glacier and decreases rapidly towards each side. Meier (1960) shows a close agreement between the measured velocity profile across Saskatchewan Glacier, Canadian Rockies, and the theoretical profile for a semi-elliptical channel.

Situations where ice flow is mainly by internal deformation, i.e. with a vertical velocity profile showing zero-displacement at the glacier sole, with a sharp velocity gradient in the basal layer and with the velocity maximum at the sur-

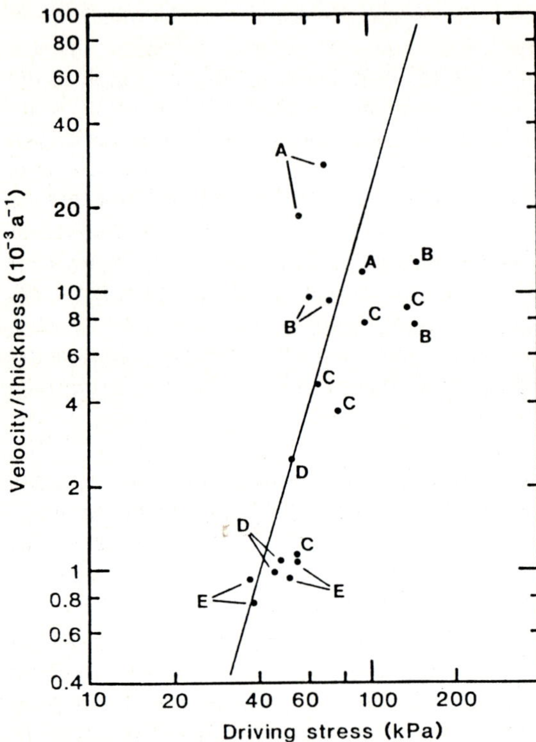

Fig. 1.4. Variations in mean shear strain rate (velocity/thickness) with driving stress upstream of a step in the Byrd Glacier profile. Regression line is for polar ice thought to be moving largely by internal deformation (Budd and Smith 1981). Letters refer to distances upstream of the step in the surface profile: *A* 10–19 km; *B* 20–39 km; *C* 40–59 km; *D* 60–79 km; *E* 80–100 km. (McIntyre 1985, Fig. 4)

face, are encountered in areas of both the Antarctic and Greenland ice sheets and in some valley glaciers. In these cases, ice is frozen to the substrate and basal temperatures are below the pressure melting point. However, in other cases, maximum calculated surface velocities achieved by creep may be significantly smaller than those actually observed. If, as indicated by Drewry (1986), h = 500 m and $\alpha = 2°$, then the deduced velocity by creep will be 50 m/year. Many glaciers exhibit average speeds of several hundred meters/year and outlet glaciers can be much faster. The difference between observed and calculated

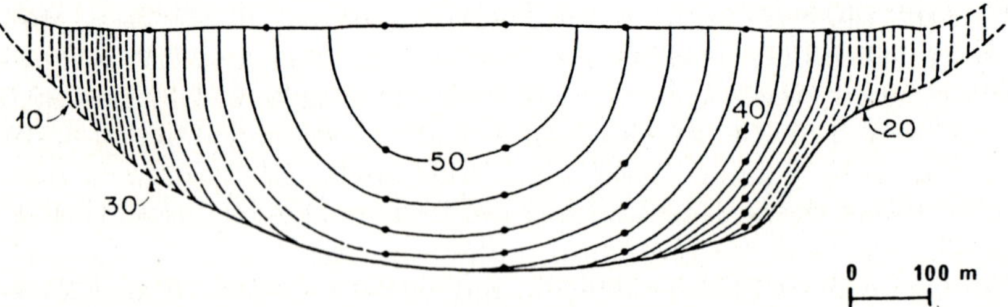

Fig. 1.5. Distribution of longitudinal velocity in a valley glacier cross section as measured in the Athabasca glacier, Canada. The units are in m/year. (Raymond 1971, Fig. 10)

velocities derived from an integration of the Glen flow law is accounted for by basal sliding.

Two processes are fundamental to basal sliding: enhanced creep and regelation. Irregularities in the bed give rise to local stress concentration on their upstream side and the creep rate is thereby enhanced. Following Weertman (1964), the smaller the size of the bed irregularities, the less is this enhancement and the lower the ice velocity. The second mechanism is dependent on the fact that if an increased stress is developed on the upglacier side of a bed protuberance, then a corresponding decrease in stress exists on the downglacier side. This gives rise to a difference in the pressure melting point at the two locations. Pressure melting on the upglacier side of the protuberance and regelation on the downglacier side explain this second sliding process. This mechanism operates if the latent heat released by regelation is conducted through the obstacle to be available for melting at the upglacier side. Effective heat transfer is difficult to sustain if the obstacle is large and the effectiveness of the regelation sliding mechanism diminishes as the obstacle size increases. The opposing nature of the two sliding processes leads Weertman (1964) to consider an intermediate obstacle size which controls the situation. This obstacle is called the controlling obstacle.

More sophisticated roughness parameters than the simple bed roughness considered by Weertman have been elaborated by Kamb (1970) and the role of basal water concentrated into cavities in the lee of obstacles has been developed by Lliboutry (1968). However, as indicated by Paterson (1981), different observations show that conditions at the glacier bed do not correspond to those assumed in current theories of sliding. No existing theory provides a realistic basal boundary condition for the glacier sliding problem.

The transition from motion by internal shear to motion by sliding frequently takes place where ice is channelled and moves over a bedrock step, often the headwall of an ice-filled fiord (McIntyre 1985). The presence of subglacial water near this point is attested to by radio-echo soundings. The decrease in surface slope occurring downglacier indicates the onset of sliding.

Until recently, glacier sliding was presumed to occur only where the basal ice is at the melting temperature, and not when it is subfreezing, because pressure-melting and regelation are necessary for sliding past the smallest scale obstacles of the bed. Shreve (1984) has pointed out that sliding at subfreezing temperatures might be possible, albeit slow, due to the presence of a liquid or liquid-like layer surrounding particles either within the ice or in subglacial sediments. Echelmeyer and Zhongxiang (1987) have made direct observation of basal sliding and deformation of basal drift at subfreezing temperatures. In a tunnel at the base of a Chinese glacier, basal sliding in the order of 0.5 mm/day was recorded at the ice-rock interface at a temperature of nearly $-5\,^{\circ}$C. This represents only a small fraction of the overall motion, which is about 15 mm/day. Indeed, two other mechanisms of sliding can account for a substantial part of the overall glacier motion: enhanced deformation of the frozen and ice-laden subglacial drift (60% of the surface velocity) and motion across discrete shear planes or shear bands within the frozen drift or at the ice-drift interface

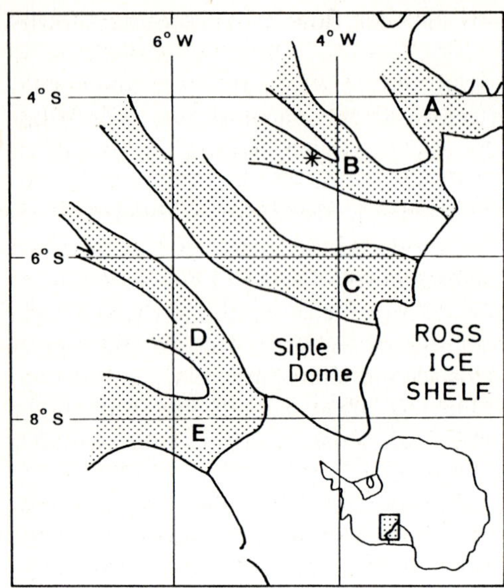

Fig. 1.6. Location map of ice streams along the Siple Coast of West Antarctica. *Stippled areas* are ice streams; the 10 km^2 measurement site on ice stream *B* is marked by an *asterisk*. (Blankenship et al. 1986, Fig. 1)

(10–20% of the surface velocity). Basal sliding at subfreezing temperatures needs to be assessed in terms of its importance in glacier flow and glacier erosion.

If a significant thickness of water separates ice from bedrock, then flow takes place through ice-bed hydrodynamic decoupling. This may be the case in ice streams. According to Drewry (1983), ice streams differ from freely floating ice shelves by the fact that water does not submerge all bed obstacles in the former, although it is present at the base in considerable amounts and gives rise to extensive decoupling. Therefore $\tau_b \neq 0$, but τ_b is significantly lower than for sliding in outlet glaciers. Under such conditions of bed separation and roughness, basal stresses could decrease with increasing sliding velocity

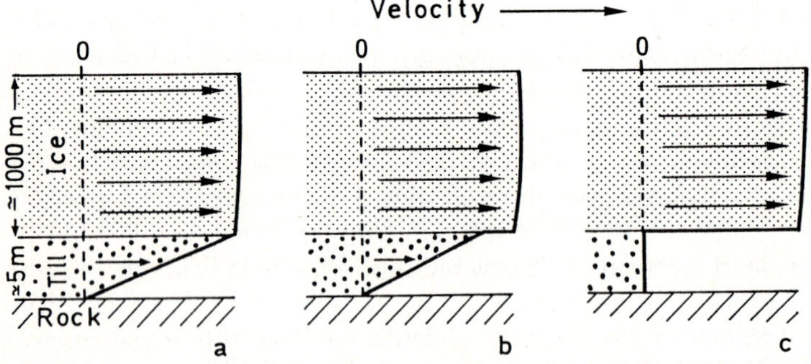

Fig. 1.7. Possible models for an ice-stream bed. **a** Till deformation only. **b** Till deformation plus basal sliding. **c** Basal sliding only. (Alley et al. 1986, Fig. 1)

(Lliboutry 1968). This is an important difference from conditions developed during "normal" sliding, where τ_b increases with the sliding velocity and where a lubrication factor enables the development of relatively high velocities after a critical limit of velocity and basal stress has been reached. Following Bindschadler (1983), the dynamics of ice streams is better explained in terms of pressurized subglacial water. High water pressures result in basal separation and in stress concentration where the ice is still in contact with the bed. It is this stress concentration that controls the sliding velocity. Such behaviour does not necessarily indicate instability; the measured velocities of the West Antarctic ice streams have been closely approximated by a steady-state approach (Weertman and Birchfield 1982). However, another factor must be taken into account. Seismic reflection studies conducted on Ice stream B (Fig. 1.6) show a 5-m-thick layer immediately beneath the ice in which both compressional (P) and shear (S) wave speeds are very low (Blankenship et al. 1986). These low wave speeds imply that the material in the layer is porous and is saturated with water at a high pore pressure. Since Alley et al. (1986) estimate that the basal shear stress is about twice the strength of this till layer, then it is considered that the layer is deforming and that the ice stream probably moves principally by such deformation (Fig. 1.7).

1.4 Ice Residence Time

Ice flow is dependent on mean annual input and output rates expressed in mm/year and per unit area of a glacier. The ratio between the quantity of ice present in a particular glacier and the input rate is the ice residence time. Let us consider two glaciers in the steady state having identical shapes and cross-sections. Let the first glacier have an average input rate per unit area of 25 mm water equivalent and let the second glacier have an average input rate per unit area of 50 mm water equivalent. In order that these two glaciers remain in equilibrium, the second glacier must transfer two times as much ice into the ablation zone as the first glacier. Since the cross sections are identical, the second glacier will have an average velocity two times greater than the first one. Ice residence time is thus two times less in the second case. The magnitude of the ice residence time can thus be known by extrapolating from average conditions at the equilibrium line and the mass loss at the equilibrium line can thereby be used to characterize the activity of a glacier.

The activity index, based on the net balance gradient at equilibrium line elevation, is expressed in mm per year per meter of altitude variation and is the sum of the rate of increase of accumulation and the rate of decline of ablation with altitude. This activity index varies between 2 and 22 mm/meter in the data of Meier and Post (1962) for ten glaciers of North America. The higher the net balance gradient, the more rapid the glacier flow will tend to be and the lower the ice residence time. There is a correlation between the magnitude of total input and the net balance gradient so that glaciers in humid areas flow faster than in dry ones. The net balance gradient is steeper in maritime climates and decreases with increasing continentality and there is a latitudinal decline in activity from temperate to polar regions. For a glacier, the oldest ice can usually be found in the frontal or marginal zone. As a result, the age of this ice is a measure of ice residence time in this particular glacier system.

1.5 Distribution of Ice Fabrics and Textures

The fabric of a crystalline mass such as glacier ice is determined by the statistical study of the orientation of the individual crystals and their relationship to each other. Orientation of the optic axis of each ice crystal in a thin section cut from an ice core is found by means of a large universal stage with four axes of rotation, mounted between crossed polaroids. A fabric diagram is obtained after plotting the optic axis of each crystal appearing in the thin section on the lower hemisphere of a Schmidt equal area projection, as is conventional among petrographers. A polycrystalline ice mass is rarely composed of randomly orientated crystals but, rather, the crystals tend to develop definite strong fabrics and systematic variations in crystal size, shape and texture.

The anisotropy, although not simple, appears to have a definite relation to the large-scale flow patterns. The main features of the large-scale deformation pattern of a glacier have been summarized by Budd (1972). Figure 1.8 illustrates the trend from longitudinal extension in the accumulation zone to compression in the ablation zone. This variation of longitudinal stress is most noticeable in the upper layers near the centre line of the glacier. Beneath the surface layers, shear parallel to the ice surface becomes dominant. Right at the base the effect of bedrock irregularities may play a major role. As a result, high longitudinal strain rates which may fluctuate greatly with such bedrock irregularities could completely destroy the effect of the horizontal shear near the bed. Anderton (1974) studied the ice fabrics in a tunnel cut at the base of Meserve Glacier in East Antarctica and showed the effect of such bed protuberances. Samples of ice cut parallel to the upper face of a large boulder sitting on the subglacial floor exhibit two maxima. The optic axes are clustered either around the vertical or perpendicular to the face of the boulder. The coarser

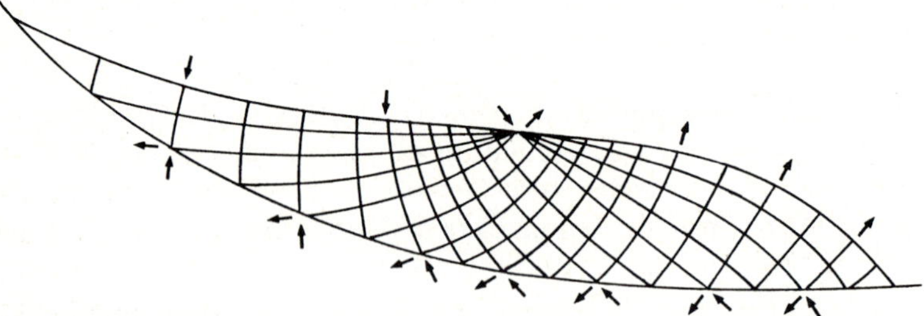

Fig. 1.8. Lines of maximal tension and of maximal compression in a viscous isothermal glacier. (Budd 1972, Fig. 1)

grains apparently retain the fabric maximum normal to the general glacier bed while the finer grains belong to the maximum perpendicular to the face of the boulder.

Lorius and Vallon (1967) conducted a detailed study of the variations in the fabric from surface to bed of a 100-m ice core from Terre Adélie in Antarctica. The fabrics from this core are shown in Fig. 1.9. At the surface, a girdle around the vertical with two opposite maxima within it predominates. This is typical of vertical compression with non-uniform horizontal extension. As the depth increases, the fabrics show a gradual change to a strong single maximum near the vertical by about half way to the bed. This is the result of the predominant influence of simple horizontal shear which prevails at depth. Further towards the base, the fabrics become much more diffuse, as one would expect

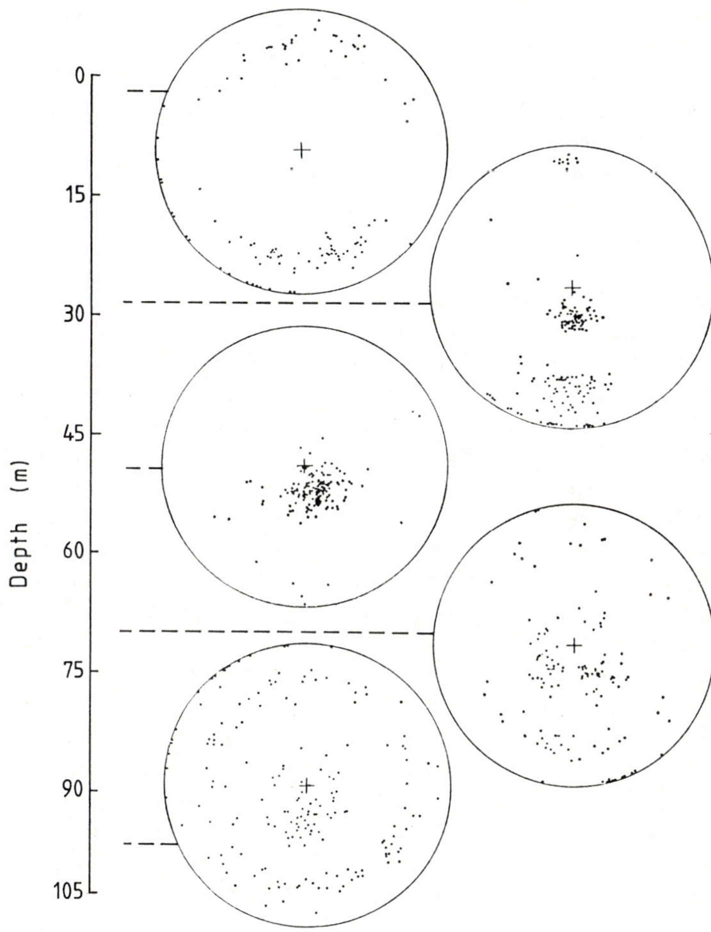

Fig. 1.9. Variation in ice fabrics from surface to bed of an ice core from Terre Adélie (Antarctica). (Lorius and Vallon 1967)

from a fluctuating stress system associated with flow over an irregular bed. The intermediate levels show gradual transitions between the three major zones discussed above.

A single vertical maximum is likely to occur in the upper layer of the ablation zone, as in Nuna ramp in North-West Greenland (Rigsby 1955). Since ice at the surface in the ablation zone represents ice that was formerly at depth further inland, this fabric may be similar to that of the zone of high horizontal shear in the vertical cores from Terre Adélie. Hooke (1970) also reported a number of ice fabrics from the ablation zone near Thule, Greenland, in which single maxima reflected the predominant internal shear. Where longitudinal strain rates are large, as near the coast in Antarctica, we may expect to obtain two maxima in the vertical plane in the direction of flow.

Since fabric development is controlled largely by the type of stress, one stress system can eventually eliminate and overprint the effect of a previous one. An ice crystal is more than an order of magnitude less viscous in a direction parallel to its basal plane than perpendicular to it (Hobbs 1974). This anisotropy causes c-axes to rotate during deformation. Ice grains that are constrained laterally by adjacent grains in a polycrystalline aggregate deform by basal glide coupled with c-axis rotation towards compressional axes and away from tensional axes. Alley (1988) indicates that, for a given bulk strain, grains that have their c-axes close to the compressional axis rotate less than grains that have c-axes at higher angles. Grain rotation will dominate fabric development until sufficient strain energy is stored to cause recrystallization, involving the nucleation and growth of new, strain-free grains. Recrystallization leads to the development of a distinctive texture of large, interpenetrating grains and to a distinctive fabric in which c-axes cluster into multiple maxima. Recrystallization is favoured by temperatures close to the melting point and large cumulative strain. It is less rapid under conditions of basal shear in ice sheets because the relative ease of deformation between adjacent grains in this case prevents large amounts of strain energy from being stored. The basal shear in ice sheets can be considered as a simple shear, in which deformation results from the shearing of a substance along a series of discrete planes.

Observations of fabrics in polar ice caps and ice sheets can readily be interpreted using Alley's model. Cold upper regions of ice sheets and ice caps, where longitudinal extension and vertical compression commonly occur, experience progressive rotation of c-axes towards the vertical without significant recrystallization. Where the shear stress is larger than the normal stress, a strong single-maximum fabric develops with a tight clustering of c-axes near the vertical. At greater depths, if the temperature is relatively high ($> -10\,^{\circ}C$), c-axis fabrics change from single-maximum fabrics to small circle girdle fabrics with multiple maxima that are centered at about 30° from the vertical. This fabric develops through recrystallization, probably in response to the combined influence of high temperature and high strain rate. At Camp Century and Dye 3 in Greenland, temperatures at the base are less than $-13\,^{\circ}C$ and a single-maximum fabric is developed. When the temperature is at the melting point or close to it, recrystallization can occur rapidly, as occurs in alpine glaciers. Small grains evolve into larger crystals up to a few cm across. Bubbles entrapped in the ice apparently inhibit the growth of crystals: many small bubbles interfere with the migration of crystal boundaries. The strain is probably relieved by migrating to a bubble-ice boundary. High temperatures approaching the melting point tend to produce larger crystals, especially where the bubble content of the ice is low. However, only large crystals have multiple-maxima fabrics. Thus, if some factor prevents crystal growth, a single-maximum fabric can persist. In the case of recrystallization at depth, debris-loaded ice near the bed can often maintain a single maximum fabric. This is probably due to the fact that recrystallization is reduced in the presence of impurities even if the thermal conditions are favourable for such recrystallization.

Crystal size distribution is an important textural parameter to consider in the glacier system. Figure 1.10 from Duval and Lorius (1980) shows crystal area expressed in mm^2 versus age at Dome C in Antarctica. This represents a situation where very low temperatures prevail (from -47 to $-53\,°C$) and where shear is negligible since the borehole at the origin of the thin sections analysed

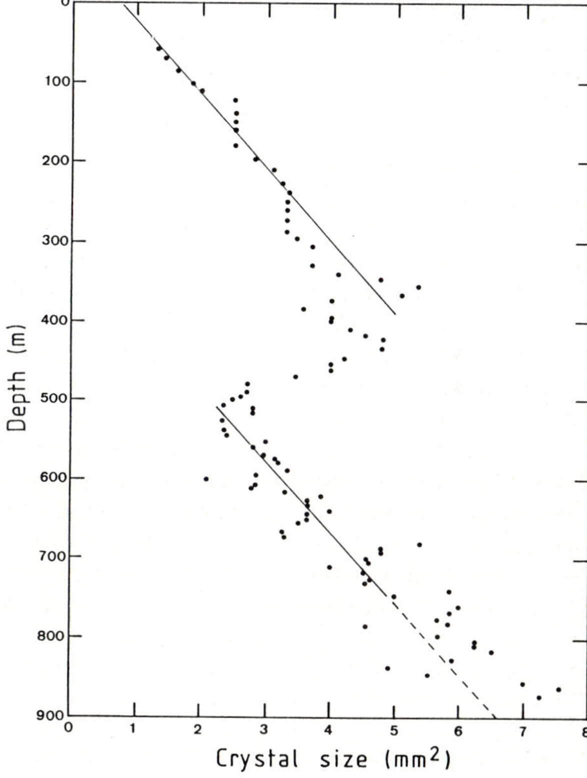

Fig. 1.10. Crystal size versus depth from the Dome C ice core (Antarctica). The *straight lines* are obtained from the linear regression of crystal size data between 60 and 360 m and between 510 and 720 m. All depths are expressed in metres of ice equivalent. (Duval and Lorius 1980, Fig. 1)

is located near an ice divide and penetrates only one-quarter of the ice thickness. Crystal mean area increases at a rate of about $3 \cdot 10^{-4}$ mm^2/year for the first 15,000 years. Then, a sharp reduction occurs when ice from the last glaciation is reached, followed by a new increase in crystal size at about the same rate as the previous one up to an age of 33,000 years at the bottom of the borehole, where the largest crystals are found (about 8 mm^2). In the Byrd Station ice core reaching bedrock at about 2400 m depth in West Antarctica, the pattern is somewhat different. After an increase to a depth of about 400 m, crystal size then remains approximately constant to 1200 m, crystal growth being inhibited by shear (Gow and Williamson 1976). A sharp decrease is observed between 1200 and 1300 m when the upper limit of ice formed during the last glaciation is reached. Crystal size then again remains approximately constant to 1800 m and finally increases by a factor of 50 between 1800 m and the bed. This represents growth by recrystallization under shear at a temperature not far from the pressure-melting point. The decrease in crystal size between Holocene ice and Pleistocene ice from the last glaciation has also been observed at Devon Island Ice Cap (Koerner and Fisher 1979) at Camp Century, Greenland, and also at Vostok, East Antarctica (Paterson 1981). Low temperatures during the glaciation, inducing a reduced crystal size or a higher concentration in microparticles which inhibits crystal growth, have been suggested as possible causes. The rheology of Holocene and Pleistocene ice will be considered later.

The presence or absence of air bubbles and their shape is another textural characteristic of glacier ice. Small air bubbles, below about 0.1 mm radius, are spherical. This is due to surface tension and diffusion. A gradient in water vapour pressure exists where there is a change in surface curvature such that vapour will diffuse from places of low curvature to places of greater curvature. Because the curvature is great in a small bubble and the diffusion path short, bubble elongation will be considerably limited in this case (Hudleston 1977). Larger bubbles tend to change shape as the ice deforms. However, they do not behave passively until they are highly elongated. Highly elongated bubbles near the base of a glacier are aligned in the plane of simple shear.

The disappearance of air bubbles can be due to meltwater giving rise to clear ice layers or to gas diffusion into the ice. The latter process explains the absence of air bubbles in the deeper part of ice sheets. For example, Gow and Williamson (1975) found no bubbles below 1100 m in the Byrd Station ice core although air was released when the ice was melted. At Vostok, in the interior of East Antarctica, there are no bubbles visible below 940 m.

1.6 Ice Foliation

In many glaciers, sedimentary bands or stratification consist of alternating layers of coarse-bubbly ice formed from winter snow and coarse-clear ice formed from refrozen meltwater. Dirt layers, formed when summer melting concentrates wind-blown dust at the surface, are also often present in sedimentary bands. This stratification is constructed in the accumulation zone but can be preserved in the ablation zone of a glacier.

Foliation is defined as a layered structure produced during ice flow under conditions of shearing or compressive strain. It consists of alternating, discontinuous layers and lenses of bubble-rich and bubble-free ice or of coarse- and fine-grained ice. As such, foliation is a secondary structure which must be distinguished from primary structures such as sedimentary bands.

Complications occur, since foliation produced during ice flow originates in some glaciers from sedimentary bands (Hambrey 1975). Indeed, foliation is usually formed by the deformation of pre-existing inhomogeneities in the ice. Crevasse traces consist of layers of coarse-grained, clear ice – frequently of dark blue appearance – which contrasts sharply with the bulk of bubbly glacier ice. They are either healed equivalents of originally open fractures or the remnants of water-filled crevasses which are now frozen. They are well developed in some Arctic glaciers like White Glacier on Axel Heiberg Island, Canada (Hambrey and Müller 1978). They can frequently be followed in the course of time throughout the length of a glacier, their curvature increasing as a result of flow. Thin sections taken through typical crevasse traces reveal that the boundary between the coarse-clear and coarse-bubbly ice is defined only by air bubble concentration, the crystals being either contiguous to those in the bubbly ice or having an identical orientation. These crevasse traces form planes that rotate from a vertical position to one that dips upglacier.

Two main stress systems can explain the main features of foliation patterns. Figure 1.11 from Hooke and Hudleston (1978) shows the deformation of a

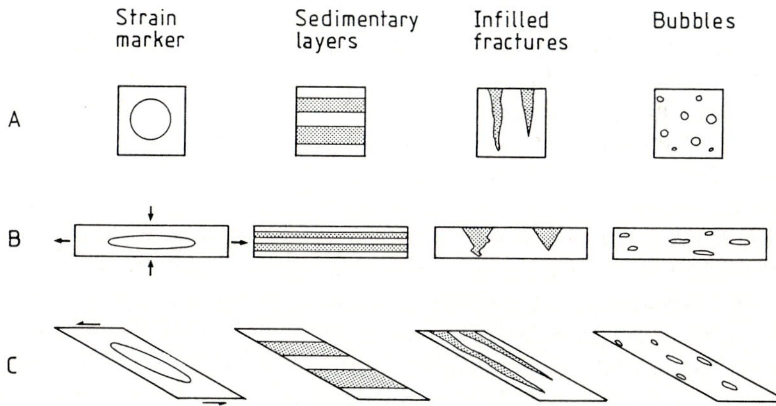

Fig. 1.11. Behaviour of various components of foliation as a result of homogeneous strain. **A** Initial configuration of components. Square and circle in left-hand diagram do not represent specific features, but are for reference only. **B** Components after pure shear with strain-ellipse axial ratio of 6.5. **C** Components after simple shear with strain-ellipse axial ratio of 6.5. The appearance of the bubbles after straining is schematic, based on observations on natural and experimentally deformed ice, and theoretical considerations. In nature the components will be affected by a complex strain history that may involve both pure and simple shear. (After Fig. 2 in Hooke and Hudleston 1978)

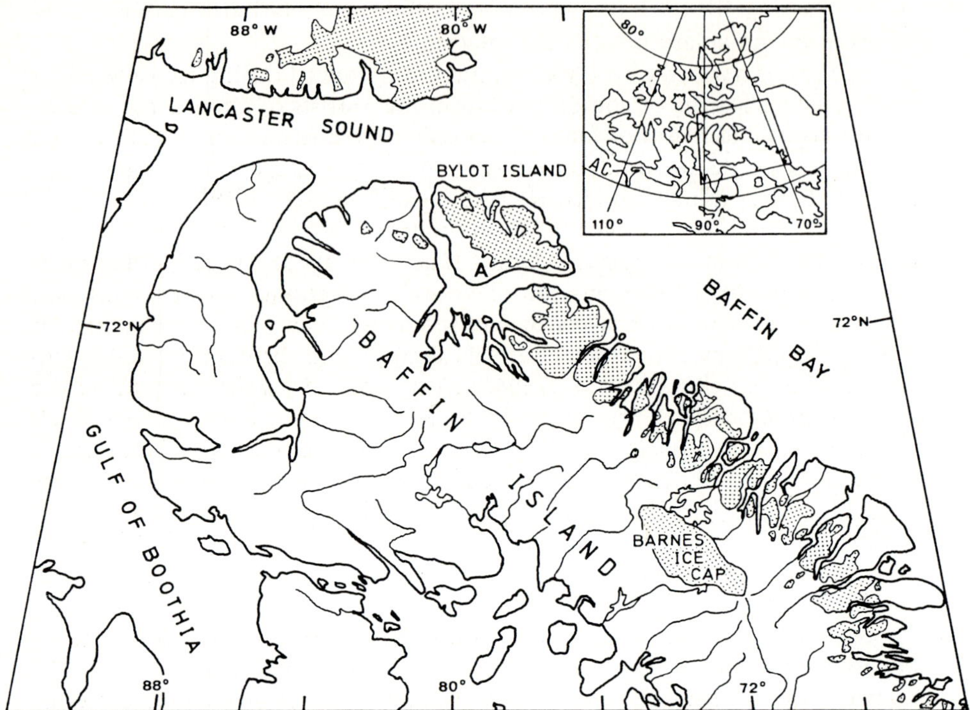

Fig. 1.12. Location map of Barnes Ice Cap, Baffin Island and of Aktineq Glacier (**A**) on Bylot Island, N.W.T., Canada. Glaciers are *stippled*

square, a circle and different components of foliation under homogeneous strain due to the two different stress systems. The first is pure shear which occurs near the surface in the accumulation zone and at all depths at the ice divide. In pure shear, a vertical compression is accompanied by an equal extension in the direction of flow. Near the surface in the ablation zone, a similar system of stresses prevails but with vertical extension and horizontal compression. The second system, which is simple shear, takes place near the glacier bed and near lateral rock outcrops where strain rates are particularly important.

Hudleston and Hooke (1980) consider that the development of ice foliation depends on the cumulative strain to which the ice has been subjected. They have computed the cumulative strains in the ice in a section through the Barnes Ice Cap, Baffin Island, Arctic Canada (Fig. 1.12). Flow transverse to the section was considered unimportant and a steady-state over the past few thousand years was assumed. Figure 1.13 illustrates the results. Flow deforms circles into ellipses. Basal ice, especially in the frontal zone, has undergone a large amount of simple shear and the ellipses are thus strongly elongated parallel to the flow lines. Sedimentary layers maintain a near horizontal attitude throughout the ice cap. Crevasse traces rotate to acquire a planar position dipping upglacier. The calculated cumulative strain is great enough to make all the stretched inho-

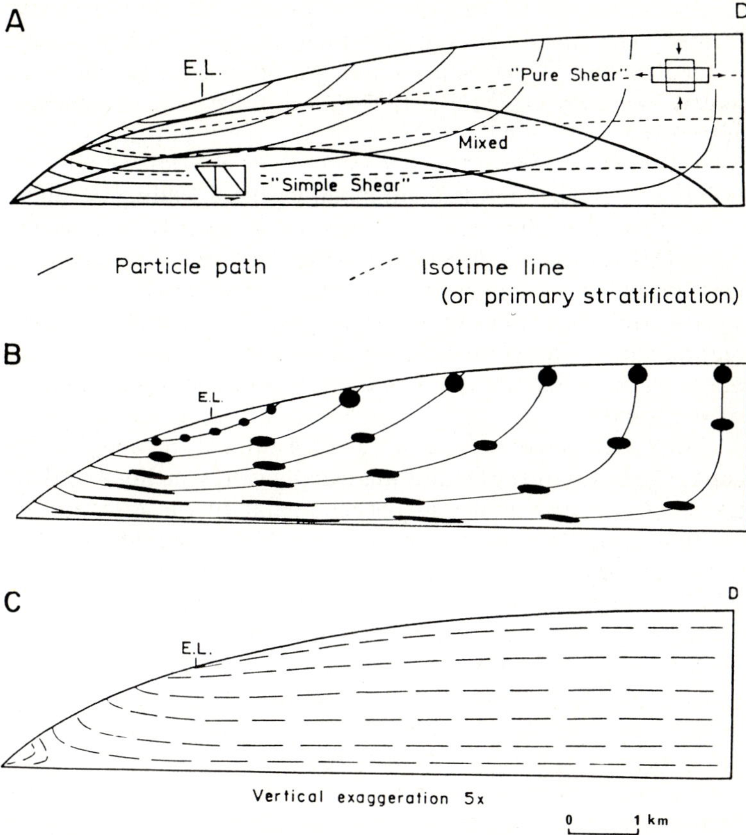

Fig. 1.13. A Schematic flow pattern in an ice sheet. *D* divide; *E. L.* equilibrium line. The strain fields delineated are for small increments of deformation only, and transverse strains are ignored. The horizontal component of the "pure shear" field will vary from extensional above the equilibrium line to compressive below. **B** Schematic illustration of cumulative strain at selected points along particle paths shown in **A**. Initial *circles* become deformed into *ellipses* during flow. **C** Schematic sketch of foliation in the south dome of Barnes Ice Cap. Pattern based on observations at surface, in a tunnel, and in several bore holes. (Hooke and Hudleston 1978, Fig. 3)

mogeneities parallel to each other near the terminus. A single foliation parallel to the flow lines thus develops. Although rotation of foliation of different origins towards a plane more or less parallel to the flow line occurs during progressive deformation, the process may not be completed if the cumulative strain is not great enough.

Longitudinal foliation, which is most strongly developed near the valley walls and roughly parallel to them, is generally believed to be the result of shear but few attempts have been made to compare developing foliation with the orientations of the principal strain-rate axes. The problem is that if rotation of the foliation has already taken place, then the measured strain rates are no longer indicative of the conditions under which the structure originally formed. The surface trace of the longitudinal foliation is essentially parallel to the flow lines. If the total cumulative strain is considered, it can be said, following Hambrey and Müller (1978), that foliation at the margins will become approximately parallel to the long axis of the strain ellipse after prolonged deformation. The same authors consider that, at White Glacier, the longitudinal foliation may have resulted from the isoclinal folding of the stratification.

Ice in cold glaciers behaves somewhat differently from ice in temperate glaciers. This is partly because the constant in Glen's flow law for polycrystalline ice depends upon temperature in such a way that the strain rate produced by a given stress decreases rapidly below 0 °C. The rate and total amount of deformation should therefore, in general, be less in cold glaciers than in temperate ones and the difference is to some extent reflected in the degree of development of the foliation.

Folded ice bands often appear in ice cliffs at the margin of glaciers and ice caps. The limbs of these folds are usually gently inclined to the horizontal and the hinge is horizontal and perpendicular to the direction of ice flow. As Hudleston (1976) suggests for the Barnes Ice Cap, folds cannot form in the foliation as long as the ice cap is in a steady state. Let us consider two particles on the same flow line. If the ice thickness changes, the particle paths will also change, depending on the subglacial topography. Let us suppose that the particle farther from the glacier margin moves to a higher flow line. Figure 1.14 is a sketch of the development of such a fold. Because velocity increases with distance above the bed, the particle farther from the margin now moves faster than the particle closer to the glacier margin. Folds with overturned limbs will form if the difference in horizontal displacement between the two particles is greater than the original horizontal distance between them. Folds are subsequently modified by shear near the bed and by compressive flow near the margin. Hudleston has found that ice samples from both limbs and the hinge of a fold displayed the same c-axes fabric, indicating that the fold was not the result of differential shear on adjacent limbs. However, the possible role of shear in the formation of some folds cannot be excluded for the moment. As the folds become tighter, the limbs can be very attenuated. This gives rise to a layered structure with an orientation parallel to the axial planes of the folds. Some banding structures also result from rotation of healed fractures. This matter will be discussed further when the basal zone of glaciers and ice caps is considered in more detail.

1.7 Basic Ice Types in Different Glacial Systems

Within the framework of the glacier system in general, different glacial systems can be defined on the basis of glacier shape. The importance of glacier shape in influencing the response of a glacier to its environment lies in the role played by the distribution of glacier area with respect to altitude. Since the net mass balance on the glacier surface varies from the head to the frontal zone, varia-

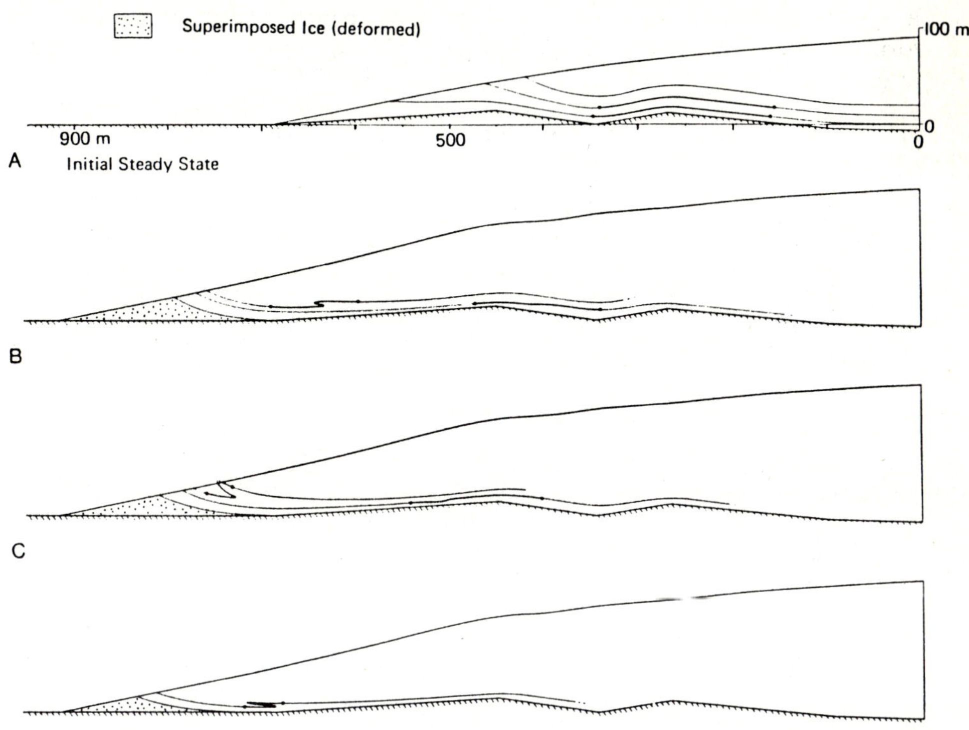

Fig. 1.14. Simple model of glacier advance and the production of folds. **A** Initial steady state: surface profile and outer 200 m of subglacial bedrock topography. Banding is assumed to have developed parallel to the three particle paths shown. **B, C** and **D** patterns of banding after a brief advance followed by renewed steady state. *Thicker lines terminated by dots* allow the flow history to be traced. Each *dot* is a material point. Upper particle path in **A** is not shown in the other diagrams. Superimposed ice was formed during the advance and became strongly deformed and overridden subsequently. (Hudleston 1976, Fig. 8)

tions in the altitudinal distribution of surface area affect the total amount of ice to be discharged. For example, large ice sheets extend for a sufficient elevation above the equilibrium line for precipitation to decrease due to altitude or the remoteness from water sources. Thus, large ice sheets have a low activity index and a high ice residence time. On the other hand, Meier et al. (1971) have noted that small cirque glaciers have particularly low ice residence times. This reflects the importance of wind-blown snow as a notable source of accumulation. This wind-blown snow originates from a wide area and accumulates in sheltered cirques so that a higher value of snow input than the regional mean gives a particularly low ice residence time.

Ice masses have been classified in a variety of ways. The widely quoted morphological classification recognizes several categories based on size, thick-

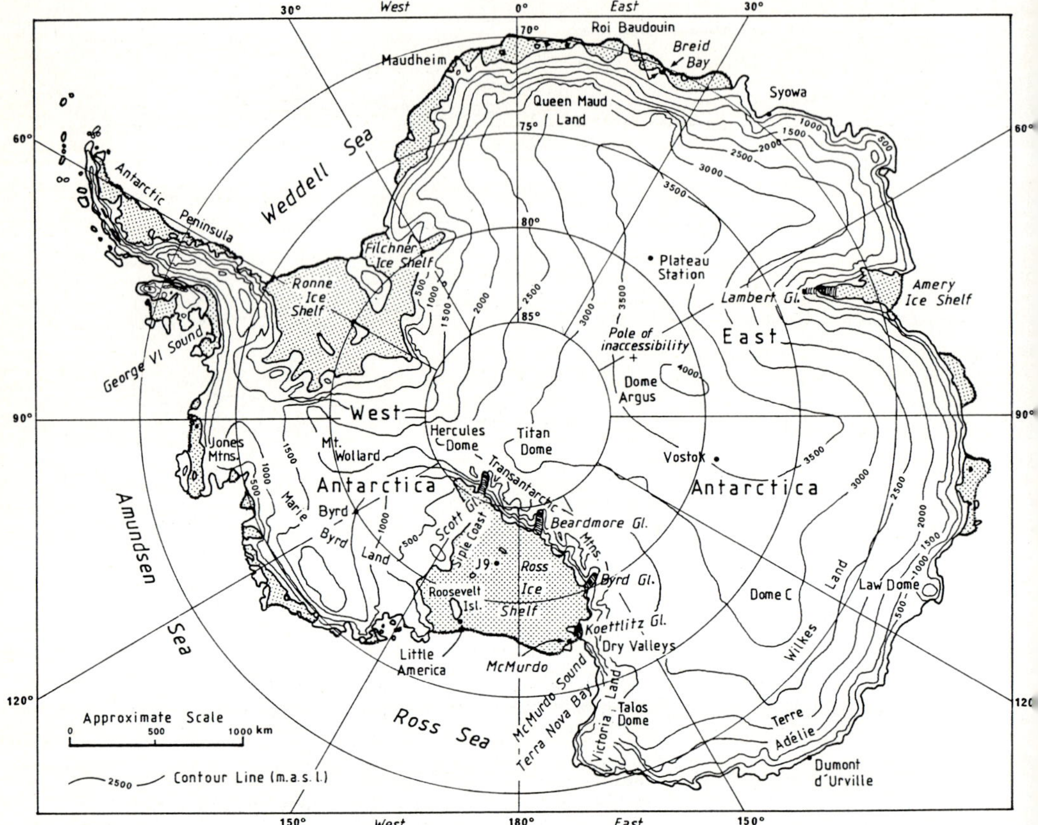

Fig. 1.15a. Location map of the main ice sheets. **a** Antarctica

ness, relationship to topography and shape. For the purpose of this book, only the following broad types will be considered:

a) *Ice Sheets and Ice Caps.* The topography is largely submerged by these ice masses. Ice movement is outward in all directions. The difference between an ice sheet and an ice cap is accepted as being one of scale (Armstrong et al. 1973), an ice sheet being over 50,000 km^2 in extent. At the present day, three such ice sheets exist: the Greenland Ice Sheet, the East Antarctic Ice Sheet and the West Antarctic Ice Sheet. These ice sheets have ice domes from which flow lines diverge (Fig. 1.15a for Antarctica and Fig. 1.15b for Greenland). Within ice sheets and ice caps some areas move faster than adjacent areas and they are called "ice streams". If, in its marginal zone, an ice sheet or an ice cap flows as a glacier tongue in a valley, then this part constitutes an "outlet glacier".
b) *Valley Glaciers.* These are completely constrained by the topography and flow in a pre-existing valley. This type of glacier is widely developed in the Alps today and, as such, will be referred to as an "alpine glacier" in this book.

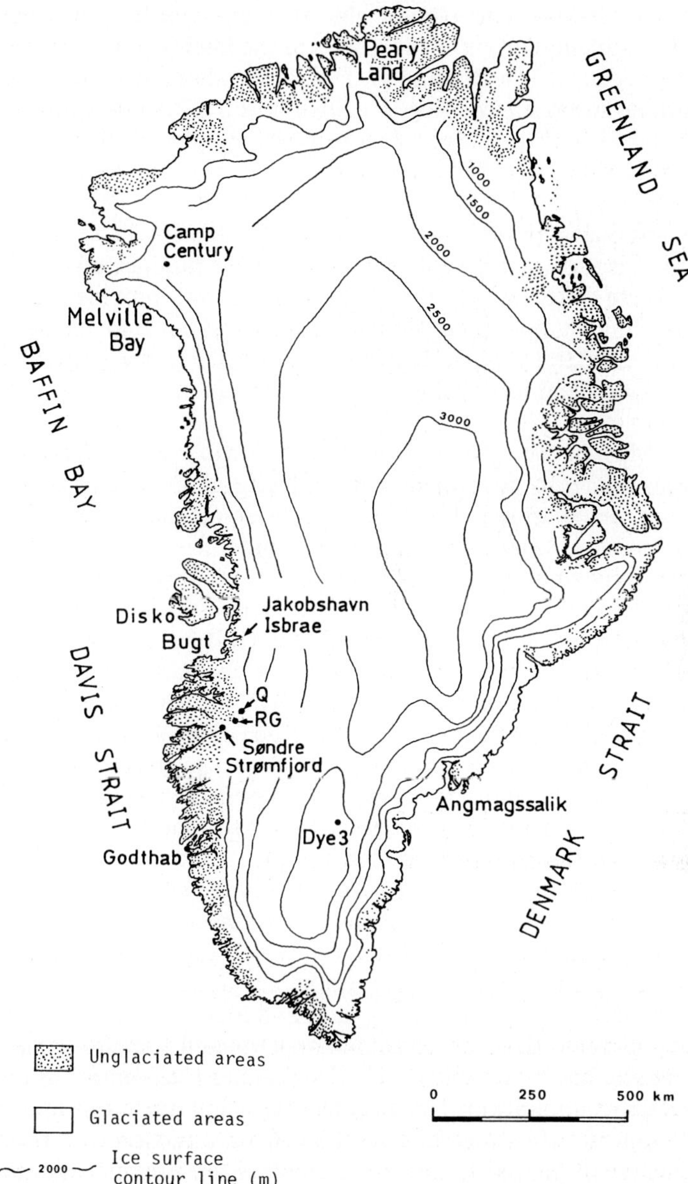

Fig. 1.15 b. Location map of the main ice sheets. **b** Greenland; *RG* Russell Glacier; *Q* Qigssertâq (see Chap. 4)

c) *Ice Shelves*. This type of glacial system can be considered as a slab of ice of continental origin, which exists at the interface between the atmosphere and the ocean. It is a floating part of an ice sheet or of an ice cap. The main ice shelves today are the Ross Ice Shelf and the Filchner-Ronne Ice Shelf, both in Antarctica (Fig. 1.15a). The seaward margin of these ice shelves generally forms a sheer ice cliff rising above sea level from which icebergs are produced.

The broad types of glaciers considered here can be viewed as separate glacial systems having different shapes. Different families of ice types are developed in these different glacial systems and they will provide the basis for an understanding of how their formation and composition may be studied in terms of their effect on glacier dynamics. A brief review of the ice types present in ice sheets and ice caps, in alpine glaciers and in ice shelves will now be considered.

Most of the surface of an ice sheet lies in the accumulation zone. In Antarctica, glacier ice is formed either in the total absence of meltwater or with only slight percolation in the uppermost layer, i.e. the dry snow zone and the percolation zone respectively. Ice formed under these conditions is characterized by relatively small (1 – 3 mm) equidimensional grains and by a great number of air inclusions. These bubbles, which are vestiges of firn pores, form a branching network, as they are located at crystal boundaries. If a small amount of meltwater has been present during ice formation, the crystals are somewhat larger (3 – 10 mm); sometimes layers of buried melt crust are encountered in cores, indicative of former brief periods of melting of the snow surface. In the Greenland Ice Sheet, more melting occurs in the lower part of the accumulation zone and superimposed ice is present at specific sites. Superimposed ice derives its physical characteristics from formation by the freezing of water infiltrating the firn pores in contact with cold ice underneath.

In superimposed ice, the size of the ice crystals is determined by the size and closeness of packing of the firn grains acting as centres of crystallization. The coarser-grained the firn from which the ice forms, the larger will be the resulting ice crystals. Large grains are about 2 – 2.5 cm, small grains about 0.05 – 0.1 cm. At low freezing rates and high water content, large crystals develop, some of them being columnar. Oriented bubbles sometimes occur if the freezing has been rapid. Since firn grains act as centres of crystallization, thin tubular bubbles will first form radially in all directions and then turn upwards or downwards, depending on the general direction of freezing. Characteristic clusters of curved filament-like bubbles are present and also large spherical, irregular or flat horizontal air inclusions which have been precipitated from the solution. Banding can occur, depending largely on the rate of meltwater freezing and the degree of saturation of the snow. Both affect the crystal size and bubbliness of ice (Koerner 1970).

Owing to the spatial migration of the different zones through time, different ice types can occur in the same vertical section. As indicated above in this chapter, these ice types undergo changes in crystal-fabric as they travel through the accumulation zone to the ablation zone.

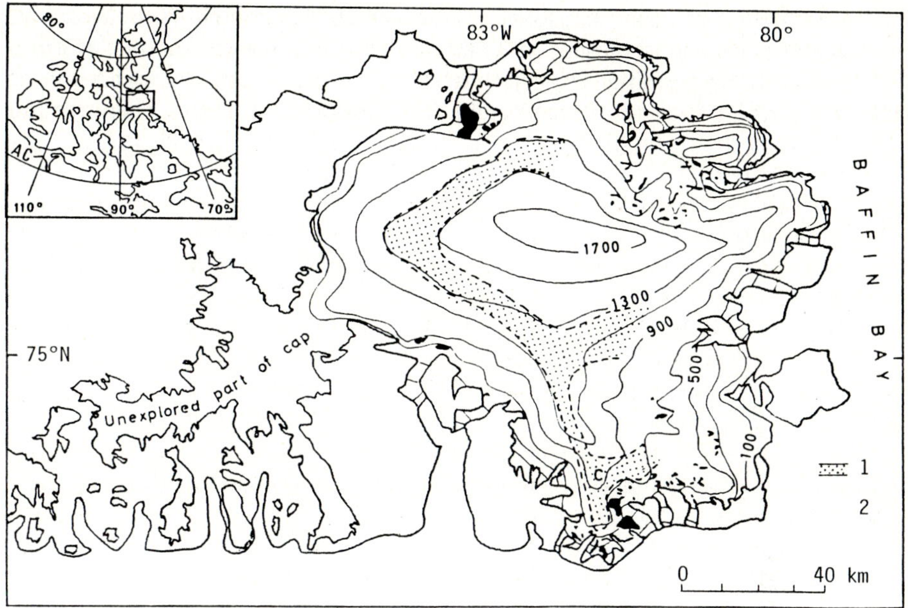

Fig. 1.16. The Devon Island Ice Cap, Northwest Territories, Canada. *1* Zone of superimposed ice. *2* Nunataks. Contour interval 200 m. (After Fig. 1 in Koerner 1970)

Towards the bed of an ice sheet, basal ice can occur. This type of ice is completely different from the glacier ice above and can usually be easily distinguished by its high debris content. Basal ice owes its origin, partially or totally, to processes operating at the base of ice sheets, ice caps or valley glaciers. It can emerge, like glacier ice, in the ablation zone of the ice sheet, deformed by the flow and the changing stress configuration.

Ice caps can also display these different ice zones, with the exception of the dry snow zone. This is the case on the Devon Island Ice Cap where most of the surface is still in the accumulation zone and where a superimposed ice zone is present (Fig. 1.16). However, in the Barnes Ice Cap, Baffin Island, also in Arctic Canada, the only type of glacier ice formed is superimposed ice and in some years the entire ice cap is in the ablation zone. Wide fluctuations in the position of the equilibrium line are thus characteristic of such an ice cap.

Valley glaciers may also exhibit different ice zones. In most cases in the Alps, ice is formed in the presence of meltwater in the firn, but superimposed ice is usually absent. White, bubbly ice with small crystals sometimes exists and testifies to the minor role of meltwater in snow metamorphism in high areas. Basal ice, although of a different nature, can often be seen along the margins of these glaciers. Foliation is often well developed in glacier ice as are crevasse fills, which constitute another ice type.

Ice shelves are nourished by snow accumulation on their flat upper surfaces and by varying amounts of ice supplied from land glaciers. Although bottom melting usually occurs at their lower surface which is in contact with the ocean, freezing at the base is present at selected sites on some ice shelves. Bottom freezing of an upper layer of ocean water is responsible for the formation of a peculiar type of ice of marine origin. The importance of this phenomenon will be stressed and developed further in the last chapter of the book.

Several types of ice are thus present in the glacier system. Due consideration of this fact must be taken into account when isotopes or impurities in ice are considered.

2 Ice Composition

2.1 Stable Isotopes and the Water Cycle

Water molecules can contain different isotopes of hydrogen and oxygen. The most abundant hydrogen isotope of mass 1 is simply designated as H. The heavy isotopes of hydrogen of masses 2 and 3 have acquired the individual names of deuterium (D) and tritium (T). This nomenclature is most often used instead of the forms of ^{2}H and ^{3}H and will also be used in this book. Tritium is unstable and decays by beta emission. For the isotopes of oxygen, there is no individual nomenclature; they are designated ^{16}O, ^{17}O and ^{18}O. The most abundant oxygen isotope is ^{16}O. ^{18}O is more abundant than ^{17}O and is often called the heavy oxygen isotope. The discussions will involve the abundance ratios R of the isotopic pairs $^{18}O/^{16}O$ and D/H.

In the commonly used δ scale, stable isotope data on natural waters are reported in terms of the ratio R between the concentrations of heavy and light isotopes ($^{18}O/^{16}O$ or D/H). The δ value of a given sample is the relative difference between the R ratio in the sample and the R ratio in a standard. The usual standard is termed SMOW or *S*tandard *M*ean *O*cean *W*ater. This is not a real water body but the zero point of the δ scale defined on the basis of a real water standard NBS-1

$$\text{So, } \delta^{18}O = \frac{^{18}O/^{16}O\ (\text{sample}) - {}^{18}O/^{16}O\ (\text{smow})}{^{18}O/^{16}O\ (\text{smow})} \times 1000 \text{ and}$$

$$\delta D = \frac{D/H\ (\text{sample}) - D/H\ (\text{smow})}{D/H\ (\text{smow})} \times 1000\ .$$

With such a definition, $\delta^{18}O$ and δD are expressed in ‰. A sample with $\delta^{18}O = +10‰$ is thus enriched in ^{18}O by 10‰ or 1% relative to standard mean ocean water. A value $\delta D = -20‰$ signifies that the sample has 20‰ less deuterium than the standard mean ocean water.

Isotopic fractionation is expressed by the equilibrium fractionation coefficient α which is defined as

$$\alpha = R_B/R_A\ ,$$

where R_A is the ratio of the heavy to the light isotope in molecule or phase A and R_B is the same in phase B. Isotopic fractionation in nature is temperature-dependent and the coefficient α generally approaches unity at increasing temperatures (Faure 1977).

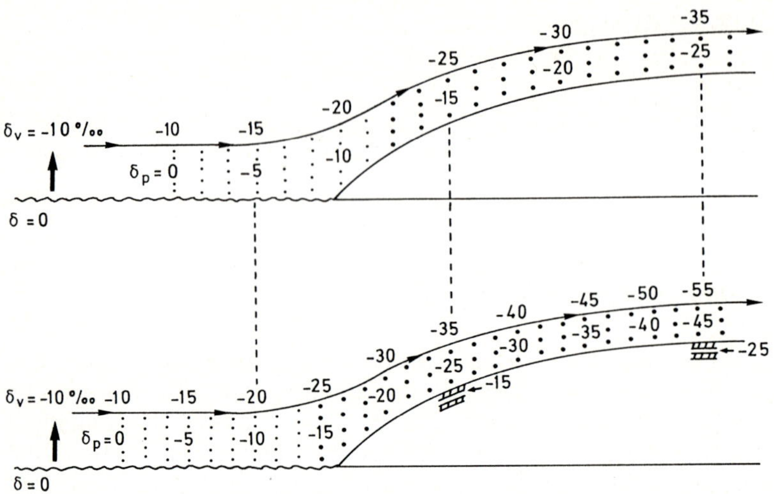

Fig. 2.1. *Upper part* (summer or warm climatic conditions): simplified circulation model showing the oxygen isotope fractionation during the evaporation of ocean water (to the *left*) and the subsequent precipitation, when the air is gradually cooled off by travelling towards higher latitudes or ascending to higher altitude over an ice sheet (to the *right*). *Lower part* (winter or cool climatic conditions): same as above, except that a cooling, increasing with latitude, has changed the isotopic fallout pattern into lower δ's at any mid and high latitude locality. Snow of $\delta p = -25‰$ is assumed to be deposited on top of snow of $\delta p = -15‰$, deposited during the preceding warm period. (Dansgaard et al. 1973, Fig. 1)

In the atmosphere, the three most important isotopic components of water – $H_2{}^{16}O$, $H_2{}^{18}O$ and HDO – are fractionated because the vapour pressure of the heavier component is slightly lower than that of the light component: about 1% lower for $H_2{}^{18}O$ compared to $H_2{}^{16}O$ and about 10% lower for HDO compared to $H_2{}^{16}O$. At equilibrium, the atmospheric water vapour will thus contain 1% (or 10‰) less ^{18}O and 10% (or 100‰) less deuterium than the mean ocean water. Consequently, the $\delta^{18}O$ and δD values of water vapour in the atmosphere above the oceans are both negative. If such a vapour is cooled, giving rise to raindrops in a cloud, the first small amount of precipitation will have the same composition as the ocean water because the vapour condenses with 10 and 100‰ preference to ^{16}O and H respectively. As shown by Fig. 2.1 taken from Dansgaard et al. (1973), further cooling leads to further depletion of both the vapour and of the condensate appearing in the later stages of the process.

The condensation of water in equilibrium with water vapour and its subsequent removal from a cloud can be described by the Rayleigh distillation equation:

$$R/R_0 = f^{(\alpha - 1)} ,$$

where R is the isotopic ratio of the remaining vapour, R_0 is the isotopic ratio of the vapour before condensation begins, f is the fraction of the vapour re-

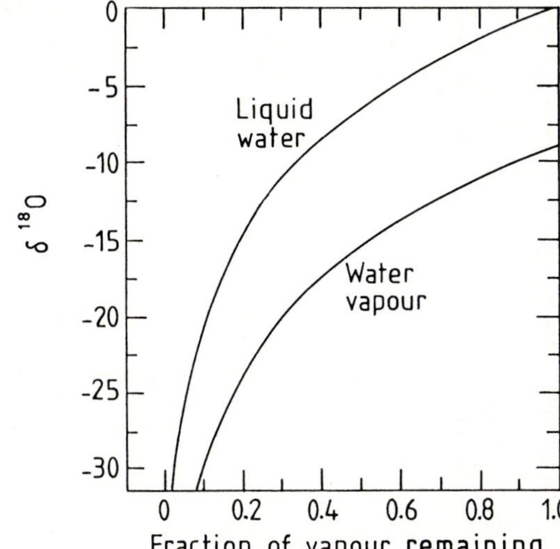

Fig. 2.2. Fractionation of oxygen isotopes during condensation of water from vapour at 25 °C, according to the Rayleigh distillation model. (After Fig. 18.2 in Faure 1977)

maining and α is the equilibrium fractionation coefficient. In Fig. 2.2 such a situation is shown for the isotopes of oxygen during condensation of water vapour at 25 °C; $\alpha(^{18}O)$ is considered to be 1.0092 and the initial $\delta^{18}O$ value of the vapour is taken to be $-9.2‰$. The first condensate in equilibrium with that vapour has $\delta^{18}O = 0$. Immediate removal of this condensate impoverishes the remaining vapour in ^{18}O and the condensate that subsequently forms in equilibrium with this vapour also acquires negative $\delta^{18}O$ values. The $\delta^{18}O$ of the condensate that is in equilibrium with the vapour at any moment can be calculated from the Rayleigh distillation equation:

$$\delta^{18}O_{liquid} = \alpha(\delta^{18}O_{vapour} + 1000) - 1000 \ .$$

A moist air mass moving towards colder regions will give rise to progressively more negative $\delta^{18}O$ and δD values of the rain falling from it. This is the result of a combination of factors. First, isotopic fractionation due to differences in vapour pressure of different water molecules at a given temperature. Second, the falling temperature of the air mass results in an increased value of the equilibrium fractionation coefficients as shown in Fig. 2.3. Third, re-evaporation of water from rain drops will impoverish the vapour phase in heavy isotopes and evapo-transpiration of water by plants will have the same effect.

Since most evaporation occurs in warm and temperate areas over the oceans, the snow that feeds the polar glaciers can be considered as produced during the last stages of the condensation process. In these stages, Dansgaard et al. (1973) consider that evaporation from solids causes no isotopic fractionation and that the exchange between ice fields and the atmosphere adds only small amounts of isotopically light vapour to precipitating air masses. This simplified model is realistic enough to explain qualitatively the spatial and

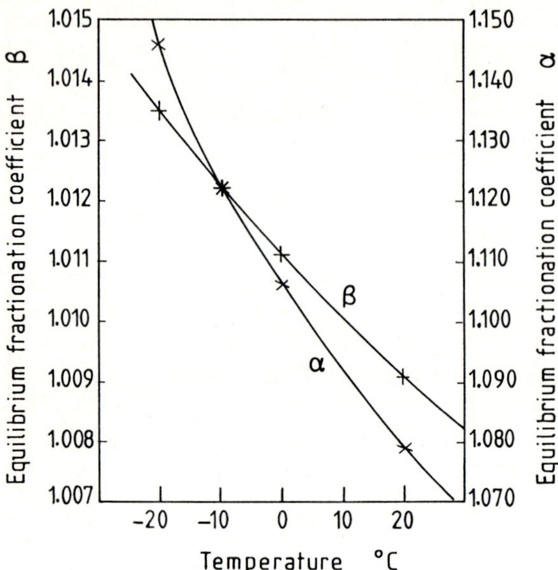

Fig. 2.3. Temperature variation of isotope fractionation coefficients for evaporation of water. The fractionation coefficient (β) is defined as the ratio $^{18}O/^{16}O$ in the liquid to $^{18}O/^{16}O$ in water vapour in equilibrium with the liquid. The fractionation coefficient for D (α) is similarly defined as the ratio of D/H in the liquid to D/H in the vapour. The graph illustrates the temperature dependence of these isotopic fractionation coefficients. (After data from Table 1 in Dansgaard 1964)

temporal isotopic variations observed in the polar regions. The preferential removal of heavy isotopes by precipitation from clouds moving towards the poles results in a latitudinal effect, giving lower δ values at higher latitudes. An altitude or inland effect is also present as cloudy air masses move up slopes. Independent of geographical δ variations, temporal δ variations have been described in the literature in terms of both a seasonal and a palaeoclimatic effect. The seasonal effect gives rise to lower δ values in winter than in summer. The palaeoclimatic effect is similarly characterized by lower δ values during cold periods as opposed to higher δ values during warmer episodes. It will be seen in the next section that kinetic effects also have to be taken into account during vapour deposition once ice crystals are formed within a cloud; the classical Rayleigh model therefore fails to explain quantitatively the deuterium and oxygen 18 contents of polar snow.

On the basis of a large number of analyses of meteoric waters collected at different latitudes, Craig (1961) has established the following major characteristics:

a) *The Precipitation Effect.* The isotopic δ values are linearly correlated over the range of variation in natural fresh waters, so that

$$\delta D = 8\ \delta^{18}O + 10.$$

The line on which the δD and the $\delta^{18}O$ values are correlated is called the meteoric water line.
b) *The Evaporation Effect.* Waters affected by a significant rate of evaporation relative to input deviate from the above relationship and show a linear correlation such that

$$\Delta D \approx 5 \Delta^{18}O \ ,$$

where Δ is the difference between the δ value of the water considered and the δ value of fresh water in the locality.

The slope of the meteoric water line is close to that which would be obtained in an equilibrium distillation column, i.e. a system in which the isotopic fractionation is determined by the vapour pressure of the different water molecules. On the other hand, the meteoric water line does not pass through the datum point of surface ocean water, as it should do in a proper equilibrium column. Undoubtedly, evaporation from the ocean surface produces vapour which is not in equilibrium with its water. The non-equilibrium character of evaporation under natural conditions has been verified by Merlivat and Coantic (1975). Subsequent precipitation, however, is assumed to be formed under conditions dominated by the equilibrium fractionation coefficient between water and vapour. The condensed phase is formed in equilibrium with the vapour and then removed from the atmosphere as soon as it is formed.

With the aim of providing readily available information on the $\delta^{18}O$-δD relationship for each station, a parameter is used which is called the deuterium excess. It has been defined by Dansgaard (1964) as:

$$d = \delta D - 8\ \delta^{18}O \ ,$$

in order to relate the isotopic composition of any water sample to the meteoric water line. For most of the stations the mean deuterium excess values are close to the global average of +10‰. However, for some stations the deuterium excess values deviate considerably from this figure. Attention must be paid not to confuse the intercept of the regression line of δD versus $\delta^{18}O$ data for any group of samples with the deuterium excess parameter d. The latter would be the intercept with the δD axis of a hypothetical line of slope 8, which is drawn through any of the points. In a way, it is more instructive to consider the deuterium excess simply as a parameter which relates the position of a point in (δD, $\delta^{18}O$) space to the meteoric water line (Yurtsever and Gat 1981). The value of d may have geophysical significance in terms of the composition of the air mass from which the precipitation derives (Jouzel et al. 1982); no such simple interpretative significance can, however, be attributed to the intercept of the δD and $\delta^{18}O$ regression lines if their slope differs appreciably from a value of 8.

2.2 Stable Isotopes in Snow

The distribution of deuterium and oxygen 18 in snow is governed by fractionation phenomena which may occur not only at the phase boundary liquid wa-

Table 2.1. Equilibrium isotopic fractionation coefficients relative to phase changes of water. (After Gat 1981, Table 4)

	α (D)	β (^{18}O)
Solid (ice)-vapour		
at 0 °C	1.132	1.0152
at −10 °C	1.151	1.0166
Liquid-vapour		
at 20 °C	1.084	1.0098
at 0 °C	1.111	1.0117
Solid (ice)-liquid		
at 0 °C	1.0208	1.0035

ter-vapour, but also at the phase boundaries ice-liquid water and ice-vapour. The respective equilibrium fractionation coefficients are presented in Table 2.1. This table shows for example that in an ice-water system in isotopic equilibrium, the δD of the ice is about 20‰ higher than the δD of the water.

The basic assumption of the Rayleigh model is that the condensed phase is formed at isotopic equilibrium with the surrounding vapour and is immediately removed from the air mass after its formation. It can be applied to liquid or solid phase formation. It must be pointed out that, because of the isotopic exchange continuously taking place between the vapour and the liquid remaining in the cloud, its use has to be restricted to clouds with low liquid water contents (Jouzel 1984). There is neither isotopic exchange with the surrounding vapour during the formation of ice crystals, nor isotopic modification of the ice crystals during their subsequent fallout to the ground. So, the hypothesis of an immediate removal of the condensate after its formation is fulfilled in this case. Polar snow formation can thus be viewed within this context. However, Jouzel and Merlivat (1984) have demonstrated the existence of an isotopic kinetic effect at vapour deposition. Once ice crystals begin to form in a cloud, vapour deposition proceeds in an environment essentially supersaturated over ice (Rogers 1979). As a result of their lower molecular diffusivity in air, HDO and $H_2{}^{18}O$ molecules tend to condense more slowly than $H_2{}^{16}O$. So, the isotopic content of the condensate is also governed by a kinetic effect and not only by the equilibrium effect resulting from the difference in saturation vapour pressures.

The existence of this kinetic effect at vapour deposition governs not only the isotopic composition of polar snow but probably also that of snow formed in mixed clouds. When there is coexistence of solid and liquid phases in clouds, the water vapour mixing ratio (g of vapour per kg of air) takes a value between that corresponding to saturation over ice and saturation over water. In this case, the droplets tend to evaporate, whereas the vapour condenses on the growing ice crystals. Both evaporation (Stewart 1975) and condensation

(Jouzel and Merlivat 1984) are accompanied by isotopic effects that include kinetic effects.

The isotopic composition of a snow cover results primarily from the isotopic composition of the snow falls of which the cover is composed. Moser and Stichler (1980) cite, in addition, two other processes: the snow cover may contain hoarfrost from water vapour in the air and snow drifted by wind from other locations. This section does not consider the transformation of the snow cover during firnification, which will be treated afterwards.

The variation in isotopic composition of successive snow layers is primarily the result of short- and long-term isotopic variations in precipitation. Picciotto et al. (1960) have shown for snow samples from the now-disappeared Roi Baudouin station in East Antarctica, that a relation exists between the δ value and the temperature measured by radiosondes in the clouds during the precipitation event. The $\delta^{18}O$-temperature gradient in East Antarctic precipitation continuously increases with decreasing temperatures, with values around 1‰ per °C at −10 °C up to 2.3‰ per °C at −50 °C. The $\delta^{18}O$-temperature relationship is thus a curved line, in contrast to the observed straight line relationships found in other polar regions by Dansgaard et al. (1973). Over Antarctica, ground and tropospheric temperatures differ, owing to the existence of a temperature inversion, the strength of which increases inland. According to Robin (1977), precipitation is formed just above the inversion layer. Condensation temperature can be estimated by the temperature of inversion. The stable-isotope composition of polar snow provides one of the most detailed proxy data for past temperatures. As indicated by Peel et al. (1988), the mean isotopic composition of snow, although primarily dependent on the mean air temperature at the deposition site, is also influenced by other factors, including distance from the source of water vapour, the rate processes of evaporation and condensation, the composition and temperature of the oceanic source, and the seasonal distribution of snow accumulation. In the Antarctic Peninsula, the δ/T gradient obtained by comparing data from closely adjacent sites, is for example some 40% less than that obtained by comparing spatial variations in the mean annual values. This arises because snowfall tends to occur when the air temperature is higher than the average for that season.

The isotopic altitude effect resulting from the temperature dependence of the equilibrium fractionation coefficients has been documented by measurements on snow samples along altitude profiles. Examples from Greenland and Antarctica are given in Fig. 2.4. This figure also shows that the isotopic composition in δD and $\delta^{18}O$ differs from one region to another not only in their absolute values but in the slopes of the linear relation between δ values and altitude. Moser and Stichler (1970) have found δD gradients varying from 2 to 6‰ per 100 m of elevation in fresh snow. It should be noted that other effects overlay the altitude effect in some areas. The lack of an altitude effect in some circumstances suggests that the snowfall is derived from a high and horizontal cloud formation (Lorius and Merlivat 1977).

Snowfalls exhibit seasonal variations of their deuterium and oxygen 18 contents. These variations remain detectable in the snow cover for a consider-

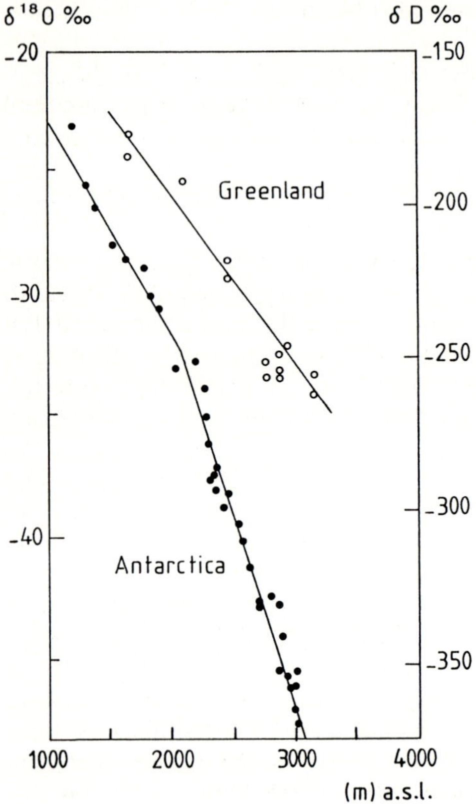

Fig. 2.4. Isotopic altitude effect in snow precipitation. The *black dots* represent deuterium measurements, the *open circles* ^{18}O measurements. (After Fig. 4.2 in Moser and Stichler 1980)

able time in the polar regions. The vertical distance between summer maxima in a detailed δ profile indicates annual net accumulation. In high accumulation areas δ oscillations with frequencies higher than one per year impede the interpretation of summer maxima, while in low accumulation areas in Antarctica the counting method can fail completely. This is indicated by the detailed study of Petit et al. (1982) in the Dome C area. These authors found fewer deuterium maxima than the estimated number of years in each dated interval, demonstrating that this deuterium profile cannot give any chronological information by itself. For dating the intervals in question, β radioactivity measurements were carried out. Changes in large scale weather patterns are also visible. For example, the summer snow layers of 1957–1958, which are relatively less impoverished in heavy isotopes in Antarctica, are traceable in the South Pole area (Picciotto et al. 1966). In Alpine glaciers, Moser and Stichler (1975) have shown that the original isotope content of the precipitation of the previous three months is preserved. They measured δD values of separate successive snowfalls on the Zugspitzplatt in the Bavarian Alps in relation to the water equivalents of the single snowfalls from September 27 to December 21, 1971. The results are indicated in Fig. 2.5. The δD values of a snow profile sampled

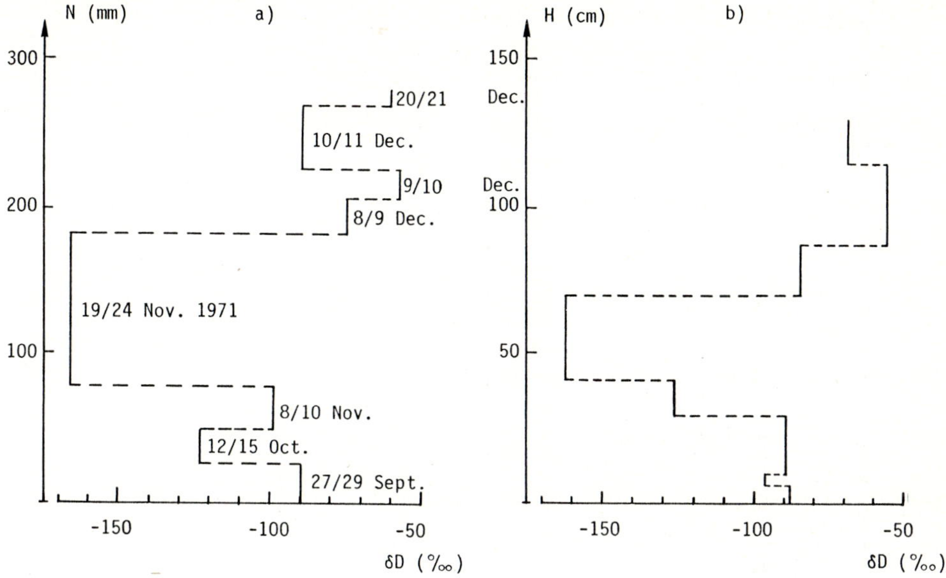

Fig. 2.5. a δD values of separate successive snowfalls on the Zugspitzplatt (Bavarian Alps) in relation to the water equivalents N (mm) of the single snow falls and respective precipitation dates. **b** δD values of a snow profile sampled on 2 January 1972, *H* (*cm*, snow height. (Moser and Stichler 1975, Fig. 7)

on January 2, 1972, plotted against height in the snow column, are indicated for comparison. At Colle Gnifetti in Switzerland, Wagenbach (1989) has indicated that the lack of annual δ^{18}O variations is primarily due to the absence of any net accumulation between autumn and early spring.

Wind drift can change the isotopic composition of the snow cover. It is particularly effective in the case of dry snow, in areas exposed to wind and with little accumulation such as in Antarctica. Wind drift can also be effective in changing the isotopic composition of snow on steep slopes where widely varying values occur. Since surface snow is dry and can be eroded, snow drift can be expected throughout the year in the dry snow zone. In other zones, snow drift occurs mainly during the cold season. Since snow drifts predominantly from higher to lower areas and the drift is most effective in removing the isotopically light winter snow, it will affect the isotopic composition in such a way that the decrease in δ values from the margin of a glacier towards the higher central parts is less than expected from meteorological data. Epstein et al. (1963) found considerably lower δ values in blowing snow at Little America in Antarctica than in any layer in a pit.

In temperate regions with heavy accumulation, snow drift is of limited importance. Ambach et al. (1968) explain the oxygen 18 content in a snowpack at the upper part of the Alpine glacier Kesselwandferner in Austria as being due to the drifting of isotopically light winter snow. It should indeed be men-

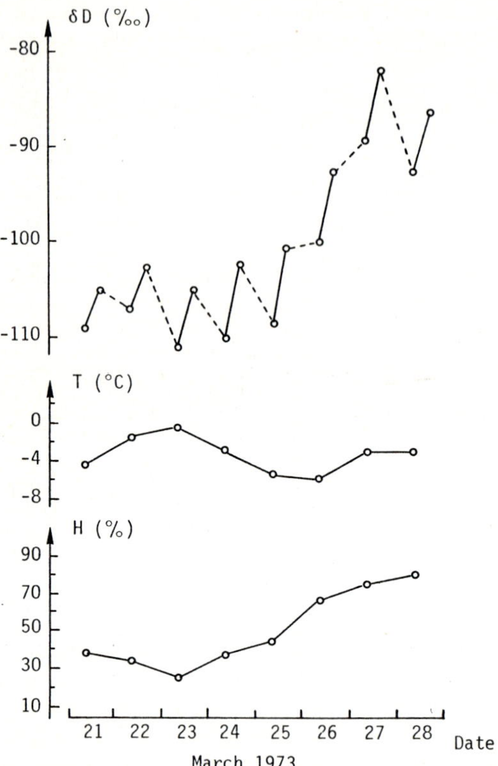

Fig. 2.6. δD values of surface snow samples, air temperature (T) and air humidity (H) from a test series at Weissfluhjoch, Davos (Switzerland). (Moser and Stichler 1975, Fig. 4)

tioned that, in these temperate regions, it is essentially the winter snow that drifts.

In contrast with a drifting of snow from higher to lower altitudes, which causes a diminution of the δ values, a snow drift to higher levels will produce higher δ values on ridges and peaks. This is connected to the formation of snow cornices.

Some isotopic effects are likely to occur when atmospheric moisture condenses on the surface of snow leading to the formation of hoarfrost. Figure 2.6 shows δD values of surface snow samples, air temperature and air humidity from a series of tests carried out at Weissfluhjoch near Davos in Switzerland in 1973 (Moser and Stichler 1975). From the curve of the δD values of the samples taken in the morning and in the afternoon, it appears that the increase in the δD value caused by evaporation of the snow at the surface is offset by condensation of air moisture during the night. However, the situation changed when new air masses with higher air moisture and different δD values were transported into the area. After a general increase of the δD values between the 26th and the 27th of March, the day-night alternation of evaporation and condensation reappeared, but occurred at a higher level.

2.3 Isotopic Changes During the Transformation of Snow into Ice

Isotopic changes occur in the first instance as a result of evaporation from the snow surface. In Fig. 2.7 from Moser and Stichler (1975), changes in δD and δ^{18}O of snow samples evaporating in a cold room at a temperature of about $-10\,°$C are given against the weight loss/g of the snow. Results show a steady increase in δD and δ^{18}O values of the remaining snow as the snow blanket undergoes progressive mass loss by evaporation. Both relationships show a strong linear correlation. On the δD-δ^{18}O diagram represented in Fig. 2.8, the same values are distributed along a straight line, the slope of which is 5.75. The snow originally introduced in the cold room is on the lower left part of the line. This effect is, however, likely to be small in nature. Hydrological and radiation energy balance studies of snow blankets indicate a commonly low value of gross evaporation rates. This is a result of the compensating effect of condensation of water vapour onto the snow surface. In the example given at the end of the previous section, the isotopic variation produced by evaporation pro-

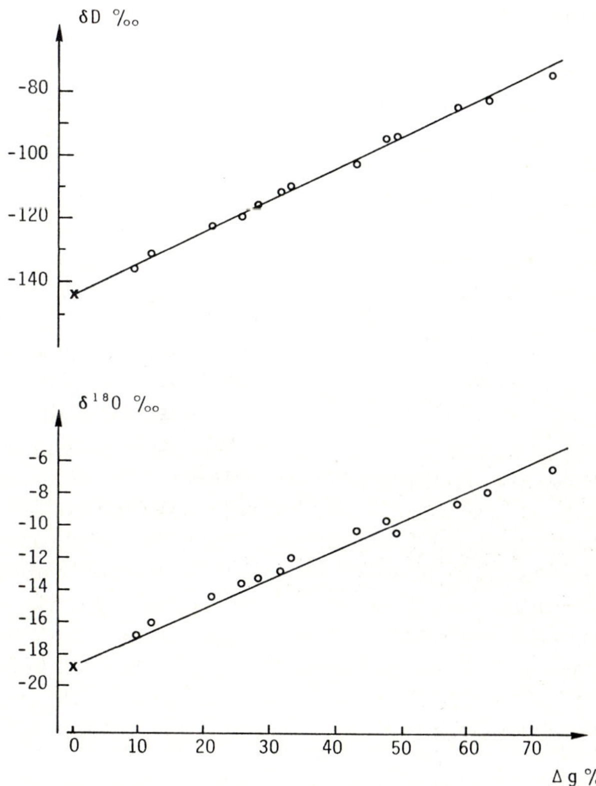

Fig. 2.7. Changes in δD, δ^{18}O values and weight loss of snow samples evaporating in a cold chamber of about $-10\,°$C. Daily net evaporation rate about 0.01 g/cm^2. (Moser and Stichler 1975, Fig. 9)

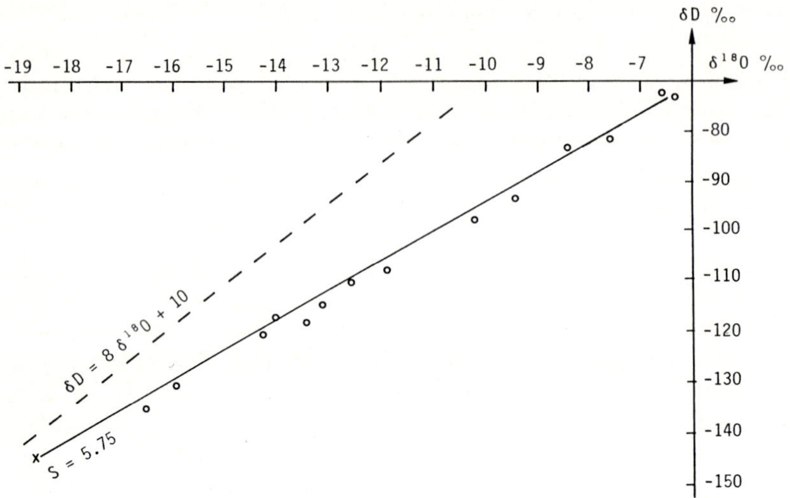

Fig. 2.8. $\delta D - \delta^{18}O$ relationship of the samples of Fig. 2.7. *S* Slope of the evaporation line. The *dashed curve* corresponds to the precipitation line. *x* Snow originally introduced. (Moser and Stichler 1975, Fig. 10)

cesses during the day is largely offset by the reverse effect due to condensation during the night.

In the absence of meltwater, the δ variations in the snow cover are reduced by mass exchange between strata via the vapour phase. According to Johnsen (1977), there are two processes by which an isotopic homogenization can occur; both based on the isotopic exchange between vapour and solid ice. They involve vertical movements of vapour as a result of changes in barometric pressure and diffusion in the vapour phase. Both these processes can cause vertical exchange in the snow and in the firn. Isotopic exchange between vapour and ice is particularly active in the uppermost part of the snow cover, where densities are still relatively low. Storms and barometric pressure changes cause vertical air movements in this upper layer, where mass exchange between the strata is further accentuated by high temperature gradients. According to Dansgaard et al. (1973), this might account for the faster smoothing observed at Byrd and Little America V stations in Antarctica which have much higher storminess than the Greenland stations. Diffusion in the vapour phase is most active in the surface layer down to some critical depth of the firn. This critical depth is, following Benson (1962), the depth where the grains in the firn become close packed. This depth is reached when the density has attained the "critical value" of 0.55 g/cm^3. At depths greater than this, isotopic exchange through the vapour phase takes place at a much slower rate until pore close-off. These two important processes for isotopic homogenization in the absence of meltwater in snow and firn cited by Johnsen (1977) are thus both connected to recrystallization of the grains via the vapour phase. The smoothing effect is dependent on the accumulation rate. In the Greenland ice sheet, an accumulation

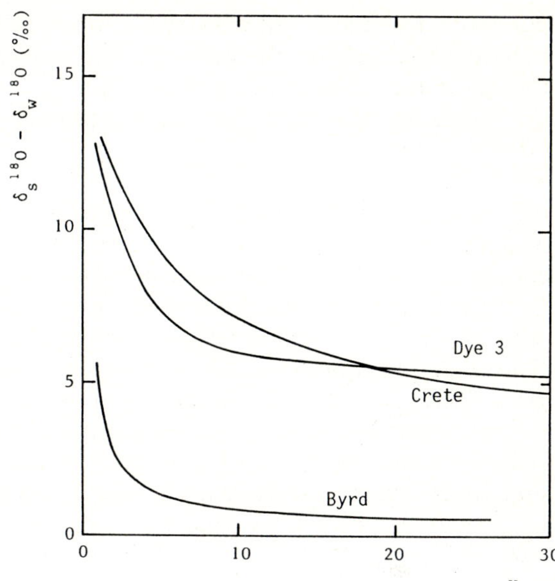

Fig. 2.9. Measured reduction of summer to winter difference, $\delta s - \delta w$, in accumulated snow as a function of time since deposition. (After Fig. 8 in Dansgaard et al. 1973)

of 24 g/cm^2/year is usually critical for the survival of the seasonal δ variation during the firnification process. Because of the storminess of the region, 34 g/cm^2/year seems to be necessary for the survival of the seasonal amplitude in West Antarctica.

Let δ_s be the δ mean value of the summer precipitation and δ_w the δ mean value of the winter precipitation. Figure 2.9 from Dansgaard et al. (1973) shows the diminution of the differences between summer and winter $\delta^{18}O$ values as a function of the time elapsed since deposition of the snow. Seasonal δ variations can be determined as long as δ_s-δ_w remains higher than 2‰. If this is the case when the firnification is completed, the stable isotopes can be used for age measurements thousands of years backwards in time. Indeed, the only remaining process of homogenization is slow molecular diffusion in the solid ice which can only be effective over very long time scales. Considering the seasonal δ variations to be harmonic implies that the precipitation is uniformly distributed throughout the year. If, for example, the winter precipitation is considerably less than the summer precipitation, the smoothing is much more effective. Similarly, individual years with unusually low accumulation will disappear in the δ profile much faster than normal years. Variations of periods shorter than one year usually disappear rapidly, but under favourable circumstances the seasonal δ variations, although strongly reduced, survive snow metamorphism and can be detected in ice cores for thousands of years. The reduction of the seasonal amplitude is inversely related to the accumulation rate. Long-term variations are not affected by these processes. If a slight melting occurs on the snow surface during a warm season, it may produce ice layers within the firn which separate the lower layers from the upper ones and prevent mass exchange due to vertical air movement. Johnsen et al. (1972) mention this as a possible cause for the unusually high seasonal amplitude observed in the Camp Century core at a depth of 776 m.

Studies on the isotopic content of a snow cover in the presence of melt- and rainwater show that the δ variations of the precipitation which are still preserved in the snowpack at the end of the winter period are usually rapidly disturbed during the following melt season. In some cases, almost complete homogenization occurs. At the same time, the remaining snow or firn becomes enriched in the heavier isotopes, deuterium and oxygen 18.

Since summer precipitation on glaciers such as those of the Alps and western Canada and United States contains much more deuterium and oxygen 18 than winter precipitation, many authors who have studied glaciers in these regions (Deutsch et al. 1966 ; Macpherson and Krouse 1967), have suggested that isotopic homogenization and enrichment in heavy isotopes are mainly the result of the freezing of rain and meltwater and the snow capture in crevasses. Free water, which usually reaches 10% by weight in a temperate snow cover, is considered to be mainly from summer rain. Sharp et al. (1960) indicate that part of the enrichment in heavy isotopes can be ascribed to free water content in the snow layers.

However, studies on Icelandic glaciers have shown that isotopic exchange during percolation may be one of the most important factors governing the homogenization processes and the enrichment in heavy isotopes. The influence of percolation of melt- and rainwater on the isotopic profile of a temperate snow cover has been studied in detail by Arnason (1969a). Iceland is especially well suited for such a study because the prevailing mild oceanic climate results in there being very little difference between the isotopic composition of winter and summer precipitation. Therefore, summer rain and meltwater produced at the snow surface contain about the same amount of deuterium and oxygen 18 as the snowpack through which they percolate. Consequently, a measured enrichment in the snow during its metamorphism could not be explained in the way suggested by the previously quoted authors.

Arnason (1981) has studied deuterium concentration versus depth in three profiles on the Vatnajökull glacier in South East Iceland (Fig. 2.10). The profiles are located at 1300, 1400 and 2000 m above sea level and extend respectively to 12, 8 and 40 m depth. These three profiles are good examples of how the degree of homogenization depends on the run-off ratio, i.e. the amount of water that percolated through the snowpack. At V-1, accumulation was estimated to be 300 g/cm^2/year. Density measurements of the firn profile indicate that approximately 50% of the precipitation escapes as run-off during the summer thaw period. At V-10, possibly 30% escapes as run-off and, at V-18,

→

Fig. 2.10. *Above* Map of Iceland showing the location of Vatnajökull where ice cores were collected. *Below* δD against depth from three pits and shallow drill holes. The *dotted vertical lines* represent the weighted mean of a given layer. The profiles of V-1, V-10 and V-18 are located at 1300, 1400 and 2000 m above sea-level respectively. (After Arnason 1981, Fig. 39)

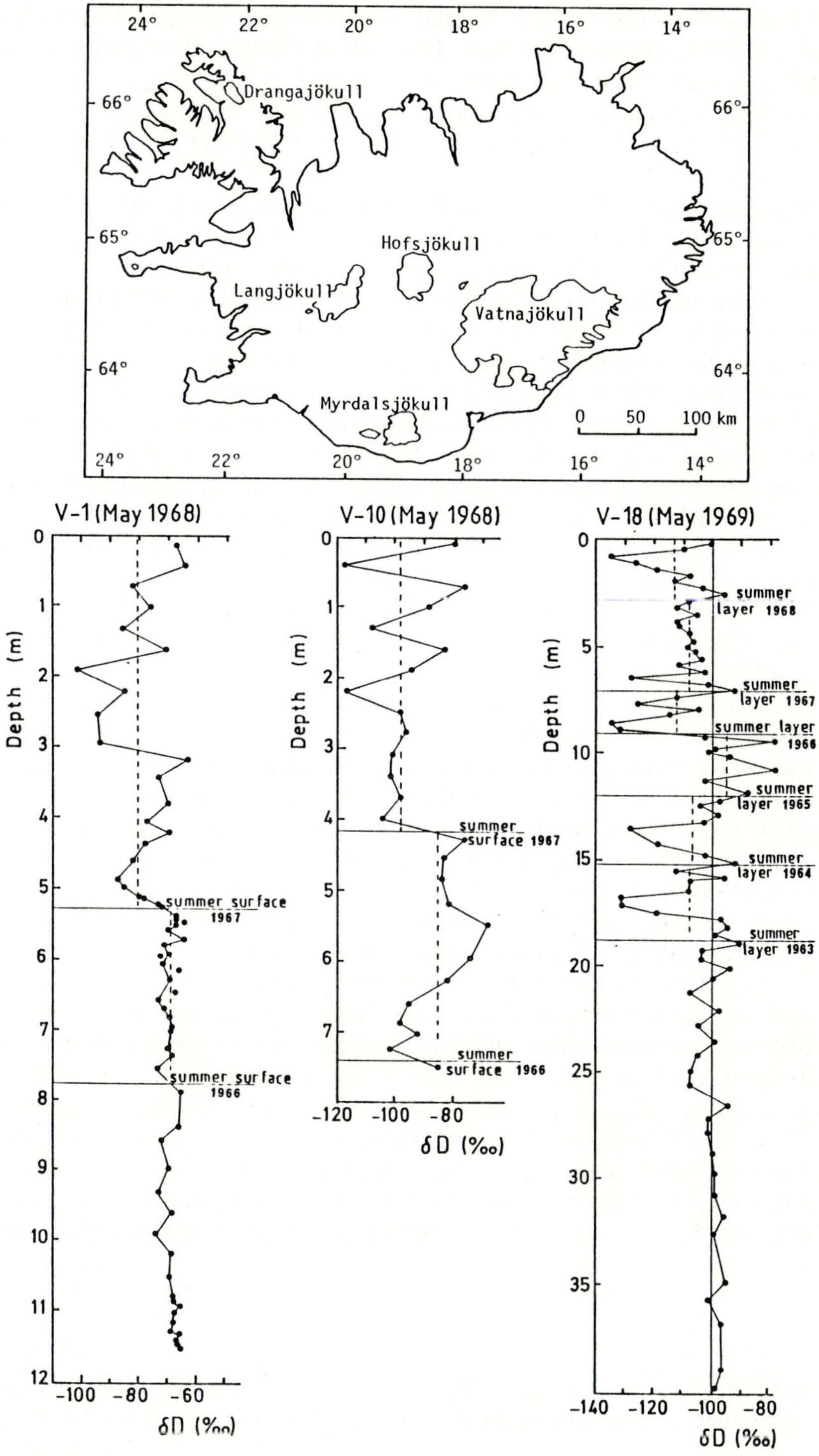

Drangajökull
Hofsjökull
Langjökull
Vatnajökull
Myrdalsjökull
0 50 100 km
V-1 (May 1968)
V-10 (May 1968)
V-18 (May 1969)
Depth (m)
summer surface 1967
summer surface 1966
summer layer 1968
summer layer 1967
summer layer 1966
summer layer 1965
summer layer 1964
summer layer 1963
δD (‰)

the amount of meltwater is so small that it does not markedly affect the initial δD variations of the snowpack. In the three profiles the δD variations within the last winter layer collected before the beginning of the melt season are of the same magnitude as the precipitation. However, below the uppermost winter layer, an extensive homogenization has occurred. An additional interesting phenomenon at V-1 is that the firn is considerably enriched in deuterium relative to the uppermost winter layer. The remaining firn has apparently been enriched in such a way that it has become richer in deuterium than even the summer precipitation. Thus the deuterium enrichment cannot be explained by the refreezing of meltwater or rain in the winter layers. The only possible explanation is that some isotopic exchange occurs as the ice recrystallizes and thus tends to reach equilibrium conditions. A comparison of the data for V-1 from below the summer surface of 1967 with those from below the summer surface of 1966 suggests that the homogenization occurred mainly during the first summer. At V-18, as indicated above, the amount of meltwater is very small. However, after 6 years the δD variations are rather rapidly smoothed out, although without noticeable changes in the mean isotopic content of the firn.

In order to understand better this exchange process, Arnason (1969a) carried out the following experiment. A column 60 cm high with an inside diameter of 3 cm was filled with dry, fine-grained snow. The snow was collected in such a way that good mixing and homogeneous deuterium concentration were ensured throughout the sample. The column was thermally insulated and melting occurred only at the top. The meltwater percolated through the snow, after which it was collected and the deuterium concentration measured. After 3 h, about two thirds of the snow was melted. The remaining snow, which had acquired a more coarse crystalline texture, was then melted and measured for its deuterium concentration. In Fig. 2.11 the deuterium content of each drain water sample is plotted against the cumulative total collected. The three points corresponding to the beginning of the experiment are scattered but all the others, obtained after the column became thoroughly wet, are aligned along a straight line. Although equilibrium conditions are not obtained in the experiment, isotopic fractionation does occur between the snow and water as the water percolates through the snow column and the remaining snow changes from fine to coarse crystalline. Considerations of the conservation of mass and isotopes give:

$$R_i \cdot q + R_w(1-q) = R_p \ ,$$

where q is the fraction of total annual precipitation remaining as ice, R_i is the mean relative deuterium concentration in the remaining ice, R_w is the mean relative deuterium concentration of the water which escapes during the summer and R_p is the relative deuterium concentration of the annual precipitation. From this, Buason (1972) was able to describe theoretically the pattern of the δD values in the meltwater, assuming in the model a partial isotopic equilibrium. Figure 2.11 shows the comparison of results from the laboratory snow melt experiment described above and calculated values from Buason. Agreement is excellent except for the first three observations. This discrepancy

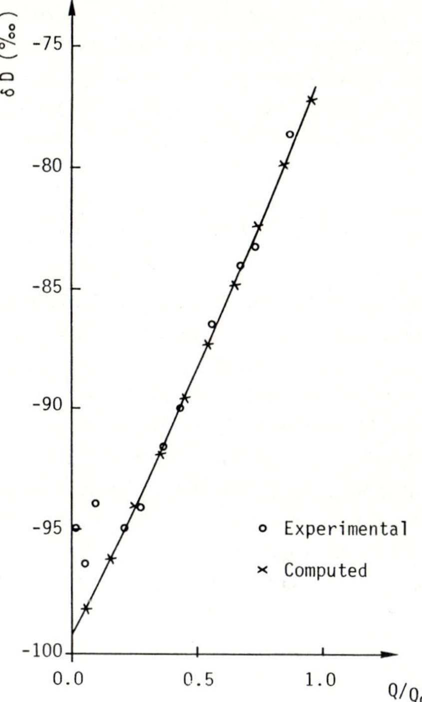

Fig. 2.11. Comparison of results from laboratory experiment and theoretical predictions in a snow percolation experiment. The δD values of drain water samples are shown versus fraction of melted snow. (Buason 1972, Fig. 9)

could have been caused by a certain portion of the first meltwater flowing down the inner surface of the glass tube instead of percolating through the snow column.

2.4 Stable Isotope Fractionation by Freezing

As the diffusion coefficients of HDO and $H_2{}^{18}O$ molecules in ice are very low – in the order of 10^{-11} cm^2 s^{-1} – melting of compact ice is likely to occur without isotopic change. Indeed, as indicated by Moser and Stichler (1980), isotope fractionation which occurs at phase boundaries is not normally observed during melting or sublimation of glacier ice, whereas it does exist if the solid phase is in the form of porous snow or firn. As seen before, there is direct evidence that an isotopic change occurs during the percolation of water through snow. Buason (1972) considers that this fractionation does not occur as a result of the phase change between solid and liquid but is probably the consequence of partial recrystallization, as indicated by a change in crystal size in the snow. The absence of fractionation during melting of ice was previously suggested by Friedman et al. (1964). It therefore seems realistic to consider that the melting of ice is not accompanied by isotopic fractionation.

Since diffusion coefficients in liquid water which can be considered as isotopically homogeneous are relatively high – in the order of 10^{-5} cm^2 s^{-1}

– freezing involves isotopic fractionation as a consequence of different water molecules freezing at slightly different temperatures. Once formed, the solid has an isotopic composition that does not change in the course of time.

Considering first a closed system, what is gained by the solid during freezing is lost by the liquid. Therefore, as demonstrated in Jouzel and Souchez (1982):

$$\delta_s = \alpha \cdot (1000+\delta_0) \cdot \left[\frac{N_0-N_s}{N_0}\right]^{\alpha-1} - 1000 \ . \quad (1)$$

In this equation, N_0 is the total number of moles in the system, or the number of moles in the liquid when freezing begins; N_s is the number of moles in the solid phase at time t; δ_s is the δ value of the solid phase near the liquid-solid interface at time t; δ_0 is the δ value of the solution when freezing begins and α is the equilibrium fractionation coefficient between solid and liquid. This equation is equivalent to that describing a Rayleigh process between vapour and liquid. It may be of interest to calculate, from Eq. (1), the mean isotopic δ value of the solid formed each time a fraction of liquid (10% for example) changes state. Let us define the frozen fraction K as N_s/N_0. From the integration of Eq. (1) between (K−0.1) and K, it follows that

$$\bar{\delta}_s = 10\ (1000+\delta_0)\ [(1.1-K)^\alpha - (1-K)^\alpha] - 1000 \ . \quad (2)$$

In Fig. 2.12, δD and δ^{18}O are plotted on a graph for ten successive fractions of ice formed during freezing of water with initial δD and δ^{18}O equal to 0 (i.e. SMOW). The values obtained are computed from Eq. (2) using $\alpha = 1.0186$ for deuterium and β (α for ^{18}O) = 1.003 for oxygen 18 (O'Neil 1968). Points for each value of the frozen fraction K are aligned on a straight line with a slope of 6.14. The value of this slope is close to the value (6.2) obtained using the ratio $S_0 = \alpha - 1/\beta - 1$, so its use as an estimate seems to be relevant. When the initial liquid is not SMOW, the slope becomes

$$S = S_0 \frac{1000+\delta_i D}{1000+\delta_i{}^{18}O} \ , \quad (3)$$

where δ_iD and $\delta_i{}^{18}$O are the δ values of initial liquid (Jouzel and Souchez 1982). On a diagram where δD is plotted against δ^{18}O, the slope on which the points representing different percentages of freezing are aligned is called the *freezing slope*. From a theoretical point of view, the value of the freezing slope must be lower if the initial water has negative δ_iD and $\delta_i{}^{18}$O values, as indicated by Eq. (3).

Souchez and Jouzel (1984) have made further theoretical developments, considering a natural reservoir with input and output, and they were able to predict freezing slopes in the case of such an open-system model. In the developed theory, a constant input, a constant freezing rate and an output proportional to the volume of remaining liquid are considered. If A, S and F are the respective coefficients for input, freezing and output and $k = N_L/N_0$ with N_L the number of moles in the liquid phase at time t and N_0 the number of moles

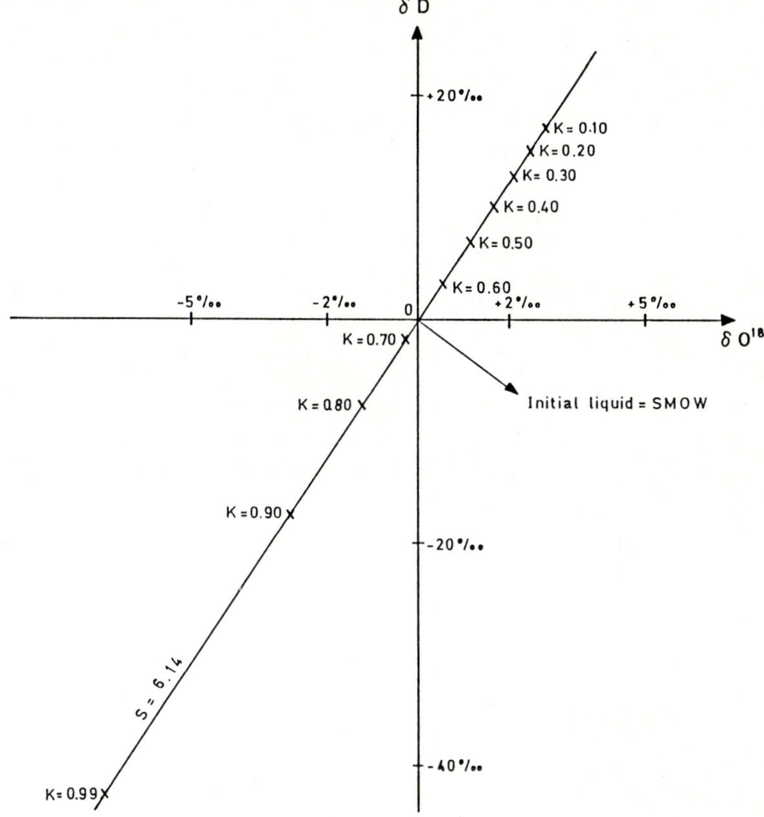

Fig. 2.12. δ values of ten successive fractions of ice formed during the freezing of SMOW. (Jouzel and Souchez 1982, Fig. 2)

in the reservoir at time t = 0, the following equation is obtained (Souchez and Jouzel 1984)

$$d\delta_L = \frac{(\alpha-1)\ (1+\delta_L)-A/S(\delta_A-\delta_L)}{k(1+k\cdot F/S-A/S)}\ . \tag{4}$$

δ_L is the δ value in the liquid and δ_A the δ value of the input. This equation, applied to deuterium and oxygen 18 using the appropriate equilibrium fractionation coefficient, gives the possibility, by a step-by-step computation, to calculate the δD and the $\delta^{18}O$ of the remaining liquid phase and of the solid formed during freezing. Results indicate that, whatever the values of F/S and A/S may be, the two phases evolve almost linearly on a δD-$\delta^{18}O$ diagram. The freezing slope can be calculated and is given by:

$$S = \frac{\alpha[(\alpha-1)\ (1+\delta_i)-A/S(\delta_A-\delta_i)]}{\beta[(\beta-1)\ (1+\Delta_i)-A/S(\Delta_A-\Delta_i)]}, \tag{5}$$

where $\delta_i = \delta_i D$, $\delta_A = \delta_A D$, $\Delta_i = \delta_i{}^{18}O$ and $\Delta_A = \delta_A{}^{18}O$. This equation shows that the slope depends on $(\delta_A - \delta_i)$ and $(\Delta_A - \Delta_i)$, the difference between the δ values of the input and of the reservoir at time $t = 0$. In a natural reservoir, there is often no reason for a change of input during the formation of this reservoir and its subsequent freezing. Therefore, in most circumstances $\delta_A = \delta_i$ and $\Delta_A = \Delta_i$ and in this case, the expression of the slope reduces to

$$S = \frac{\alpha[(\alpha-1)\ (1+\delta_i)]}{\beta[(\beta-1)\ (1+\Delta_i)]}. \tag{6}$$

Equation (6) is independent of A, F and S and almost identical to the equation obtained for the closed system. The theory thus predicts that the freezing slope will be the same for an open system as for the closed system when the input is not significantly different in its isotopic composition from that of the natural reservoir being considered.

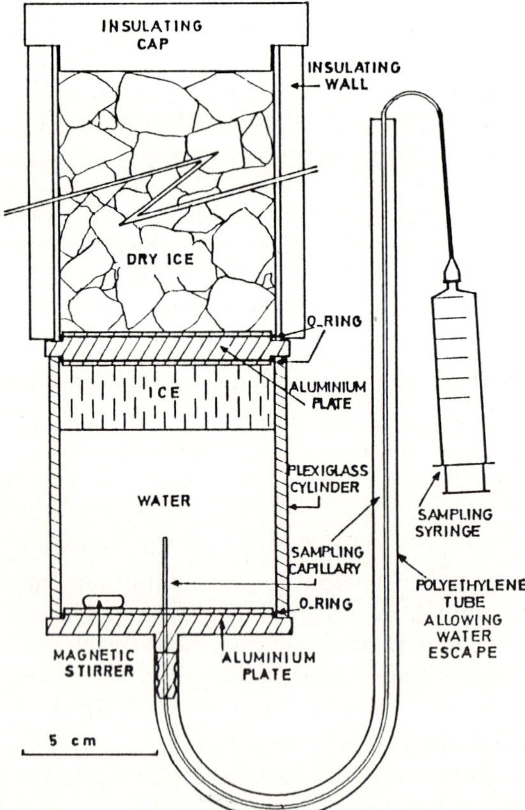

Fig. 2.13. The experimental vessel used for unidirectional freezing of water. (Souchez and Jouzel 1984, Fig. 1)

This model has been tested against experimental results. Experiments on progressive unidirectional freezing have allowed the determination of the changes in δD and $\delta^{18}O$ of the successively-formed ice layers and the corresponding remaining waters. In each experiment, 500 ml of water was progressively frozen downwards in a Plexiglas cylinder approximately 10 cm high and with an internal diameter of 8 cm, shown in Fig. 2.13. Excess water produced during freezing was diverted through an adjacent tube and allowed to escape. The residual water was continuously stirred during freezing with a small magnetic stirrer, reasons for which will be explained later. The freezing front moved downwards as a well-defined macroscopic plane. During each experiment, one millilitre of water was collected five times using a capillary tube inserted through the adjacent tube into the remaining water present in the cylinder. Residual water was sampled at the beginning of the experiment and after 40%, 80%, 85%, 90% and 95% of the liquid had frozen. When the freezing process was complete, the ice core was recovered and sectioned. The samples covered the freezing ranges from 0 to 10%, 10 to 20%, 20 to 50%, 50 to 60%, 60 to 70%, 70 to 80%, 80 to 90% and above 90% for the last slice. The ice slices were allowed to melt completely before being transferred, in the same way as the water samples, into glass bottles. The samples were analysed for δD and $\delta^{18}O$ in the same aliquot by twin mass spectrometers designed for simultaneous analysis (Hagemann and Lohez 1978). Precision of the measurements was ±0.5‰ for δD and ±0.1‰ for $\delta^{18}O$.

Three experiments were carried out with waters having very different initial isotopic compositions: meltwater from an Antarctic ice core with an initial δ value for deuterium (δ_i) of −408.8‰ and an initial δ value for oxygen 18 (Δ_i) of −51.70‰ (graph A in Fig. 2.14), water from a small pond on Victoria Island in the Canadian Arctic with $\delta_i = -153$‰ and $\Delta_i = -19.10$‰ (graph B in Fig. 2.14), and water from a small pond near Grubengletscher, Switzerland, with $\delta_i = -109.1$‰ and $\Delta_i = -15.25$‰ (graph C in Fig. 2.14). Results of the experiments are indicated in Fig. 2.14. The numbers quoted increase from the beginning of the experiment to the end: from 1 to 6 for the water samples and from 7 to 14 for the ice layers. Three main points arise from this figure:

a) A progressive impoverishment of the heavy isotopes in both the residual water and the ice layers is clearly visible during the course of freezing.
b) Water and ice samples lie on a straight line. Using a least-squares method, the slopes are respectively $S = 4.37 \pm 0.11$ for the Antarctic meltwater, $S = 5.99 \pm 0.10$ for the Arctic water, and $S = 6.63 \pm 0.17$ for the Alpine water. The correlation coefficients are all greater than 0.996.
c) The freezing slope for the Antarctic melt water is lower than that for the Arctic water, which is lower than that for the Alpine water, in accordance with the initial isotopic composition of the samples.

If Eq. (6) is used with $\alpha = 1.0208$ (Arnason 1969b) and $\beta = 1.003$ (O'Neil 1968), the calculated slopes are respectively 4.32 for the Antarctic meltwater, 5.99 for the Arctic water and 6.27 for the Alpine water. Experimental slopes are thus well in accordance with those predicted from the model.

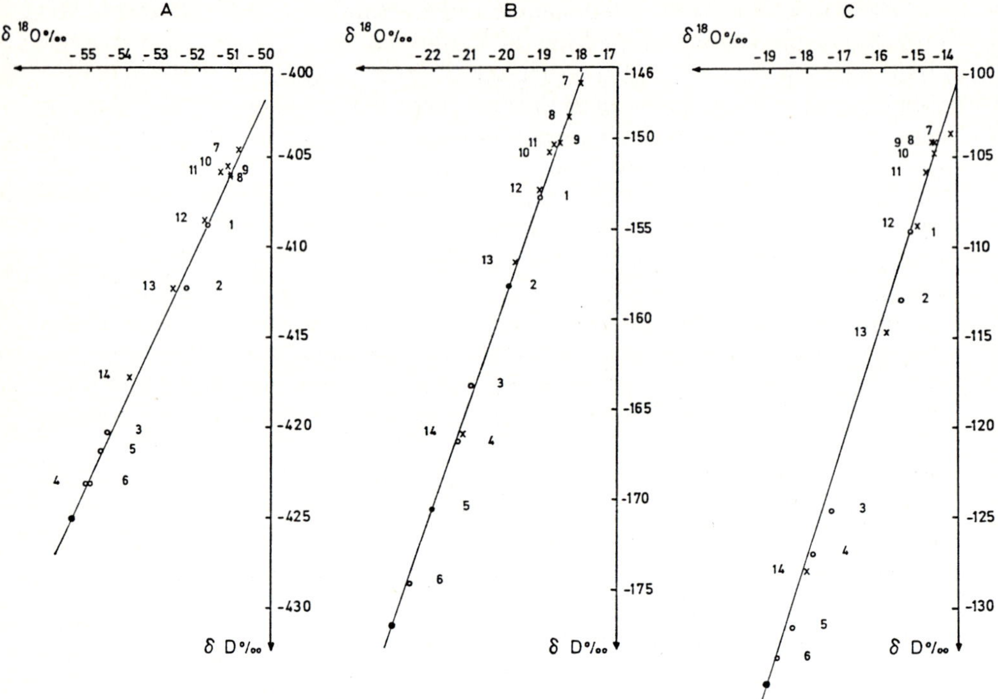

Fig. 2.14. The isotopic composition of water (*1* to *6*) and ice (*7* to *14*) during freezing experiments **A, B** and **C** (see text). The *straight lines* are the calculated slopes from the model. The *black dots* on the lines are the calculated ice values at 95% freezing. Water samples are denoted by *circles*, ice samples by *crosses*. (Souchez and Jouzel 1984, Fig. 2)

The open-system model developed above assumes a constant freezing rate but this is not the case in the experiments. However, the experimental results do not show a change of slope in the course of freezing in spite of a continuous reduction in the freezing rate. Moreover, the introduction to the model of a time-dependent freezing rate will make the computation more complicated but will have no effect on the slope predicted. If attention is now focussed on the δ range between the ice (or residual water) at 95% freezing and the initial value, this is less in the experiments than that predicted by the model. The observed enrichment of the ice compared with the bulk of the liquid water depends on several parameters, particularly on the freezing rate and on the possible trapping of liquid during crystal growth. This dependence between the freezing rate and the observed fractionation is discussed in Posey and Smith (1957), O'Neil (1968) and Arnason (1969b). As noted by Posey and Smith (1957), the observed separation will be less than the true value and the effect can be reduced by strong agitation and very slow freezing rates. Previous experiments without stirring led to correct slope values but to an even smaller range and thus to lower precision. In Fig. 2.15 the apparent $(\alpha-1)$ values are plotted

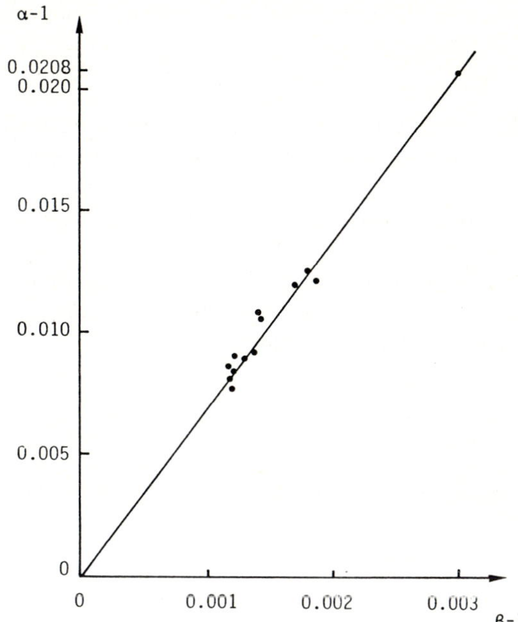

Fig. 2.15. Relationship between apparent fractionation coefficients in the freezing experiments of Fig. 2.14. Equilibrium values are respectively: $\alpha_{eq}-1 = 0.0208$ and $\beta_{eq}-1 = 0.003$. (Souchez and Jouzel 1984, Fig. 3)

versus the apparent $(\beta-1)$ values in the three experiments for different liquid fractions. A straight line joins the zero point where there is no fractionation to the point where $(\alpha-1)$ and $(\beta-1)$ have their equilibrium values. From this figure it is clear that the points are close to the straight line indicating that the slopes $(\alpha_{app}-1)/(\beta_{app}-1)$ in the experiments are not different from the equilibrium slope $(\alpha-1)/(\beta-1)$. It is quite probable that, for freezing rates encountered in nature, the slope S is always a characteristic of the freezing process, even if apparent fractionation coefficients have values lower than their respective equilibrium values. The trapping of liquid water which subsequently freezes during crystal growth will lower the range but will have no effect on the slope, since residual water samples and ice samples lie on the same straight line on a δD-δ^{18}O diagram.

The freezing slopes in open and closed systems are practically the same if the isotopic composition of the input in the open system is not significantly different from that of the initial reservoir. A computer program has been devised to simulate the evolution in δD and δ^{18}O of ice samples if the reservoir allowed to freeze is mixed in the course of time with an input having a lighter isotopic composition than the initial liquid (Souchez and De Groote 1985). This has been made by a step-by-step procedure using Eq. (4). In this program the freezing factor $K = 1-k = 1-(N_L/N_0)$ varies from 0.01 to 0.99 in 99 steps and the ratio between the input rate coefficient A and the freezing rate coefficient S, from 0.1 to 0.9. For each value of A/S, a line joining 99 points is obtained. The isotopic composition of the initial reservoir has been fixed at

$\delta D = -118‰$ and $\delta^{18}O = -16‰$. The values are related to each other by the equation

$$\delta D = 8\ \delta^{18}O + 10\ .$$

With an input having the same isotopic composition, results of the simulation give, for different values of A/S, superimposed lines forming a freezing slope that corresponds exactly to the one calculated from Eq. (3). Figure 2.16 gives the results of the simulation with an input having $\delta D = -138‰$ and $\delta^{18}O = -18.5‰$. These values are also on the meteoric water line with the deuterium excess equal to 10. No output has been considered. In this figure (and in the next one), the diagonal joining the upper right-hand corner to the lower left has the equation

$$\delta D = 8\ \delta^{18}O + 10\ .$$

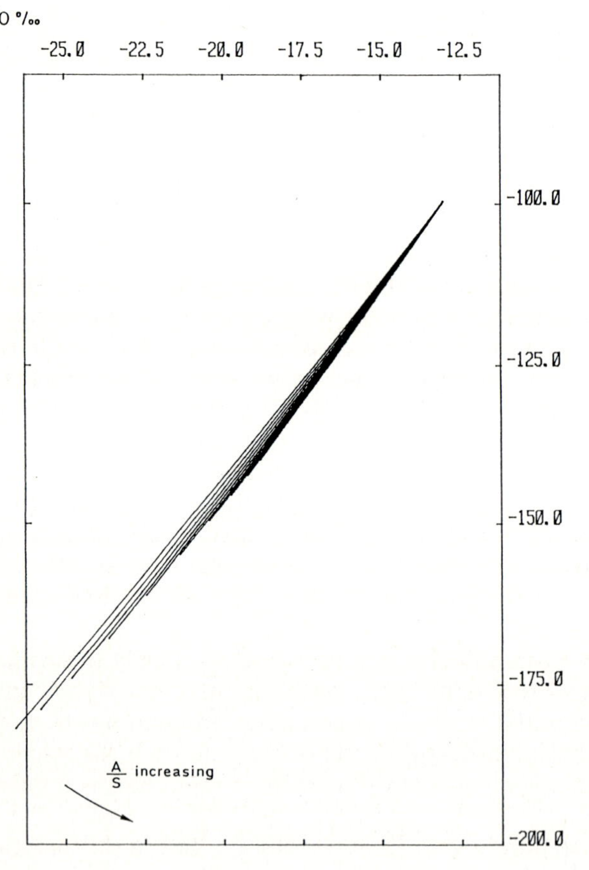

Fig. 2.16. Computer simulation of the evolution in δD and $\delta^{18}O$ of ice during freezing if an initial reservoir is mixed with an input having a lighter isotopic composition. (Souchez and De Groote 1985, Fig. 3)

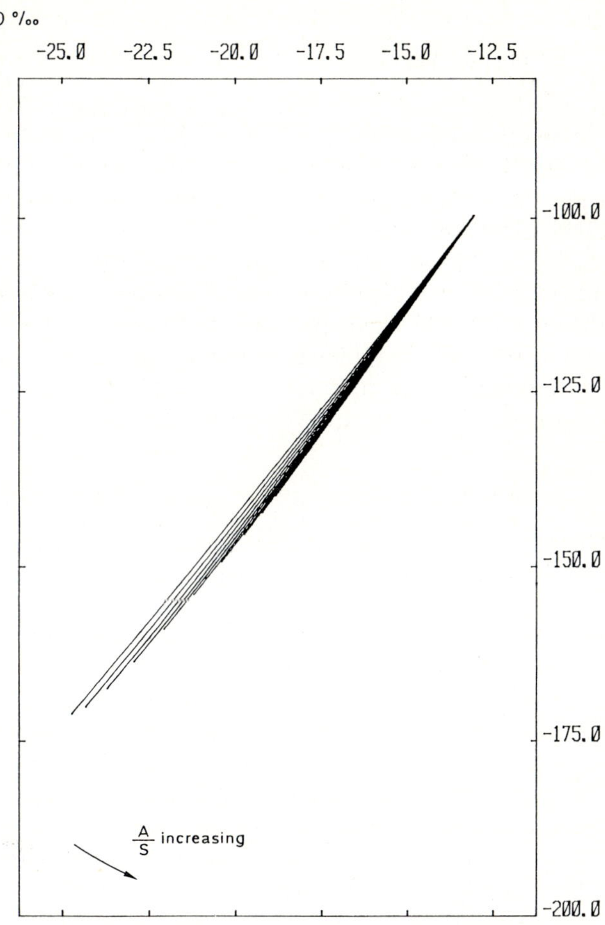

Fig. 2.17. Computer simulation with the same conditions as in Fig. 2.16 but with an output. (Souchez and De Groote 1985, Fig. 4)

Results indicate a progressive change of slope around a pin point at $\delta D = -97.2‰$ and $\delta^{18}O = -13.0‰$. If A/S increases, the slope becomes steeper until it approaches the value of 8 but with a lower deuterium excess. Figure 2.17 shows a similar pattern for a situation with the same input and with an output F = 0.9 S, having an isotopic composition identical to that of the reservoir. If, instead of the previous range of variation, A/S varies from 1.1 to 10.1, the same trend is also observed. Thus, the computer simulation indicates that the open-system model, with an input poorer in heavy isotopes than the initial reservoir, will lead to a progressive change of slope around a pin-point that can be calculated from the equation

$$\delta D = \delta_i D + 20.8$$

and

$$\delta^{18}O = \delta_i^{18}O + 3 \quad ,$$

if $\alpha = 1.0208$ and $\beta = 1.003$. Outflow of water from the reservoir will not change the general trend.

Distribution of a heavy isotopic species in ice during water freezing is related to the distribution in the liquid immediately adjacent to the freezing front. Since diffusion in ice is slow ($\approx 10^{-11}$ cm^2 s^{-1}), the distribution produced in the ice can be obtained from knowledge of the compositional variations in the liquid at the interface. These variations depend on mixing, that occurs by diffusion and by convection. The first case to consider is that of sufficiently strong mixing to maintain a uniform concentration throughout the water at all times during unidirectional freezing. This situation of complete mixing homogenizes the whole liquid reservoir at all times. A Rayleigh distribution is obtained in the ice:

$$1 + \delta_s = \alpha(1 + \delta_0)f^{\alpha - 1} \quad ,$$

where $1 + \delta_s = R_s$ is the isotopic ratio in ice, $1 + \delta_0 = R_0$ is the isotopic ratio in the initial water, α is the equilibrium fractionation coefficient and f is the remaining liquid fraction. Such a situation, which occurs for very low freezing rates, is indicated in Fig. 2.18a. The complete mixing is due to diffusion and/or convection. Another limiting case is that in which transport within water occurs by diffusion only and a concentration gradient exists in the reservoir ahead of the freezing front. Such a situation is displayed in Fig. 2.18b. Ice has an initial isotopic ratio $R_s = \alpha\ R_0$. Equilibrium fractionation always occurs at the interface and water is depleted in the heavy isotope, thus establishing a concentration gradient and inducing diffusion of the heavy isotope from the main body of water. In this situation, a steady state may occur in which the input into the diffusion layer is equal to the output into the ice. The region in which the isotopic ratio in ice falls from its initial value of αR_0 to the steady state R_0 is termed the "initial transient". A "terminal transient" occurs when the ice-water interface approaches the end of the reservoir such that the diffusion necessary to maintain steady-state freezing is no longer possible.

Although mixing can be achieved throughout the bulk of the liquid, a zone in which transport takes place only by diffusion always exists as a boundary layer of thickness BLT, adjacent to the ice-water interface. Because of the existence of this boundary layer, the uniform concentration throughout all the water is not attained and this will affect the distribution of isotopes in the ice. The thickness of this boundary layer depends upon the amount of mixing present in the liquid. The isotopic distribution in the ice is similar to that for the case of complete mixing except for an initial transient. This pattern is the most probable in natural environments. The general distribution of an isotopic species in ice formed by the migration of a well-defined planar freezing front through water is thus affected by several parameters such as the initial water δ value, the diffusion coefficient, the rate of freezing, the length of the reservoir and the boundary layer thickness. The effect of these parameters will now be studied.

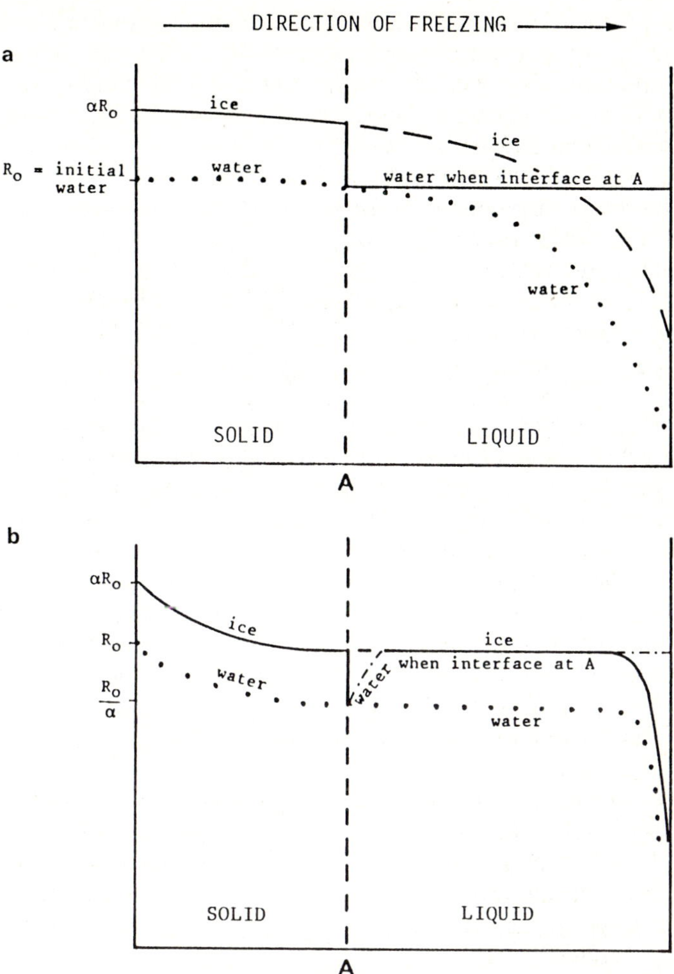

Fig. 2.18. Isotopic distribution in ice and water during unidirectional water freezing. R_0 is the R ratio of initial water. **a** Complete mixing occurs through diffusion and/or convection and homogenizes the whole reservoir at all time (Rayleigh distribution in the ice). **b** Mixing occurs through diffusion only and a concentration gradient exists in the reservoir ahead of the freezing front

Modelling the redistribution of isotopic species during water freezing is mathematically equivalent to modelling the redistribution of a solute during solidification, a problem extensively studied for the preparation of materials by zone melting. For the one-dimensional case in which growth proceeds by the movement of a planar interface separating liquid and solid and, in the absence of convection, this problem has been solved analytically for simple diffusion (Tiller et al. 1953; Smith et al. 1955). Burton et al. (1953) and Wilcox (1964) introduce the concept of the boundary layer. Such an approach is quite complex if the freezing rate varies during the freezing phase.

Souchez et al. (1987) develop a numerical approach using a simple box diffusion model, introducing the constraint that pure diffusion is limited to the boundary layer and considering that complete mixing occurs in the remaining part of the reservoir. Results of the computing with different values of BLT, for a given freezing rate of 0.5 cm/hour and a total length of 10 cm, are shown in Fig. 2.19a. If BLT = 0, a Rayleigh distribution is displayed. For BLT ≥ 1 cm, the curve is identical to the one obtained by a simple diffusion process. For lower values of BLT, a part of the initial transient exists, followed by a Rayleigh-type distribution with an apparent fractionation coefficient different from the equilibrium value. The part of the initial transient common to all the curves is very well approximated by:

$$\delta_s = \delta_0 + (1+\delta_0)(\alpha - 1)\,e^{-(\alpha V/D)x} \quad , \tag{7}$$

where x is the frozen thickness, V is the freezing rate, D is the diffusion coefficient, δ_0 is the initial δ value of the liquid and δ_s is the δ value of the ice. Equation (7) is similar to the one used in solute freezing. The difference lies in the fact that selective incorporation of isotopes is considered instead of preferential rejection of solutes.

The curves in Fig. 2.19b are computed for a fixed BLT of 0.08 cm at freezing rates ranging from 0.1 to 2 cm/h with a reservoir length of 10 cm. The initial transient is different for each curve and it is clearly seen that, for higher freezing rates, diffusion in the boundary layer is less efficient to compensate the depletion of heavy isotopes at the interface, the δ value in the ice being lower at a given distance in the reservoir. Above 2 cm/h in this case, a steady state is displayed.

Freezing rates are usually dependent on the thickness of the ice already formed, i.e. the position of the interface at time t. Applying heat transfer theory and experimental work, Terwilliger and Dizio (1970) have found that, for a fixed temperature at the cold end, the interface position x at time t is given by

$$x = (2\tau t)^{1/2} \quad ,$$

where t is the time elapsed since the beginning of freezing and τ is a constant that can be found from the slope of a linear least squares plot of x versus $t^{1/2}$. The freezing rate is thus related to the increase of the interface position by

$$V = \frac{\Delta x}{\Delta t} = \frac{\tau}{x} \quad .$$

Introducing these varying freezing rates to the model described above, a δD distribution curve in ice is obtained for given τ and BLT values (Fig. 2.19c). This curve shows an initial drop, a minimum and a reverse gradient before the Rayleigh-type decrease. For increasing values of BLT, the position of the minimum is changed and the reverse gradient decreases. For a given reservoir, once the position of the minimum and the value of the reverse gradient are determined, τ and BLT are known.

In order to test the model experimentally, three unidirectional freezing experiments were carried out in a cylinder similar to the one represented in

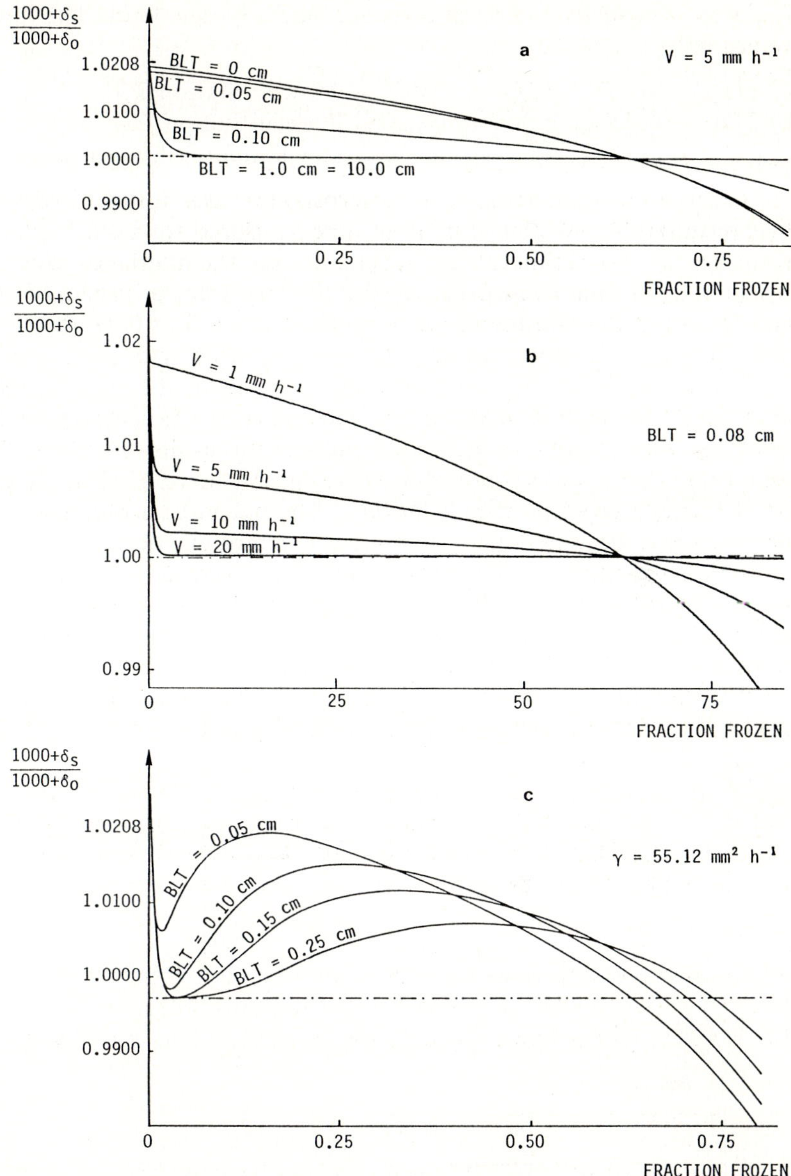

Fig. 2.19. Results from the computer simulations of the δ distribution in ice against the fraction frozen (finite reservoir). **a** Effect of the boundary layer thickness at a given freezing rate (V = 5 mm/h). **b** Effect of the freezing rate with a given boundary layer thickness (BLT = 0.08 cm). **c** Effect of the boundary layer thickness for a decreasing freezing rate with a given value of τ. δ_s and δ_0 are the isotopic compositions (in deuterium) in the ice and in the initial water respectively. (After Souchez et al. 1987, Fig. 1)

Fig. 2.13 with meltwater from different Antarctic ice cores. The initial water values were:

a) $\delta D = -282.1‰$ $\delta^{18}O = -34.65‰$ Plateau Station PS1
b) $\delta D = -425.9‰$ $\delta^{18}O = -54.20‰$ Plateau Station PS2
c) $\delta D = -159.2‰$ $\delta^{18}O = -18.30‰$ Roi Baudouin Station RB1

The water was not agitated and a cryostat was used instead of dry ice. Interface positions in the course of time were measured for each freezing experiment: results show a linear relationship between the interface position and the square root of time in accordance with the heat transfer model of Terwilliger and Dizio (1970). At a minimum confidence interval of 0.99, the slope of the best fit line is identical in the three experiments and τ can be taken as 55.12 mm^2/h. From this, the hyperbolic decrease of the freezing rate can be calculated. Taking this decrease into account in the simulation of the model described above, with the appropriate initial water values, Souchez et al. (1987) were able to obtain a theoretical δD distribution curve. Figure 2.20 shows this curve for each experiment in solid lines. The ice cylinder obtained in each experiment was sliced in a cold room using a microtome device allowing a 2-μm resolution and slices of approximately 150 μm were sampled and melted for isotopic analysis. Results of the isotopic analyses in δD are plotted on Fig. 2.20

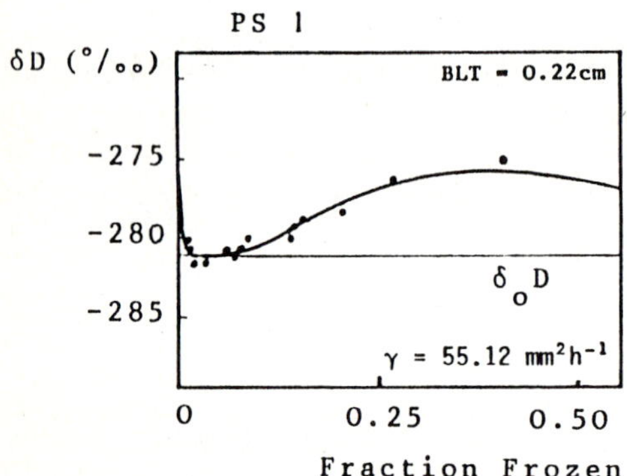

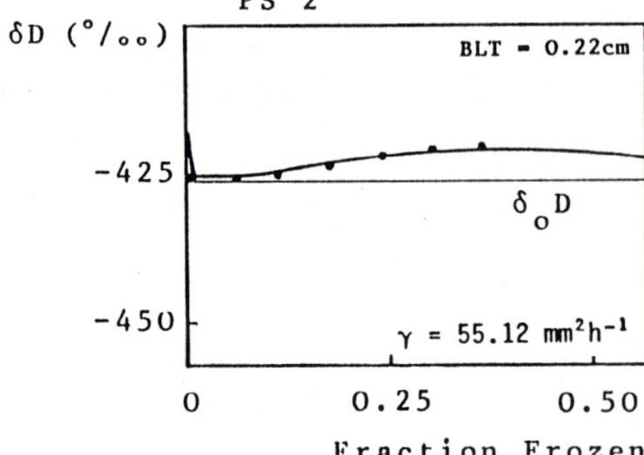

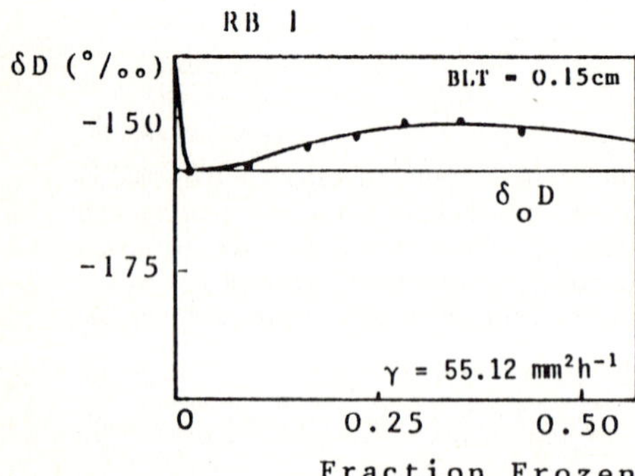

Fig. 2.20. Theoretical curve (*solid line*) and experimental results (*black dots*) for the δD distribution in ice in three freezing experiments. (After Fig. 2 in Souchez et al. 1987)

as black dots. In the three experiments, the minimum and the reverse gradient are observed and, for PS1, where more values were obtained at the onset of freezing, the initial drop is displayed. A close agreement exists between model and experiment, indicating a possible use of the theory for determining freezing rates in nature.

The theory of freezing rate prediction based on the isotopic composition of the ice and developed above can therefore be applied if the position of the freezing front at the beginning of freezing is known. It implies the determination of the isotopic profile over the whole thickness of the ice formed, requiring the sampling of successive thin ice layers. This is quite feasible as the deuterium or the oxygen 18 content can be measured with an accuracy of 0.5‰ or 0.1‰ in very small samples (0.3 ml). It is likely that the potential of the technique will diminish at lower freezing rates as one approaches a situation of equilibrium fractionation.

Now, when the length of the reservoir is more than one order of magnitude larger than the thickness of ice formed, one can consider that the δD or δ^{18}O value of the bulk reservoir remains constant during freezing and equal to the initial water value. Figure 2.21 shows the model results in this case. The initial transient is still present, but the Rayleigh-type distribution is replaced by a steady state at a given δD or δ^{18}O value depending on the freezing rate. If the freezing rate changes during the freezing process, a shift will occur toward another steady-state isotopic value in the ice, in accordance with the new freezing rate. In this particular case, the relationship between the freezing rate and the isotopic composition in the ice can be very well approximated by an equation similar to Burton's equation (Burton et al. 1953):

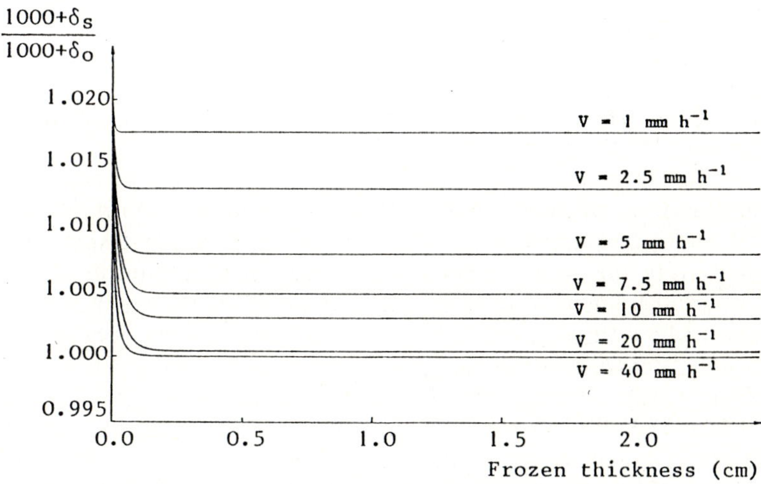

Fig. 2.21. Effect of the freezing rate with a given boundary layer thickness (BLT = 0.08 cm) in the case of a semi-infinite reservoir. The isotopic distribution displays a steady state at a level depending on the freezing rate. δ_s and δ_0 are the isotopic values (in deuterium) in the ice and in the initial water respectively

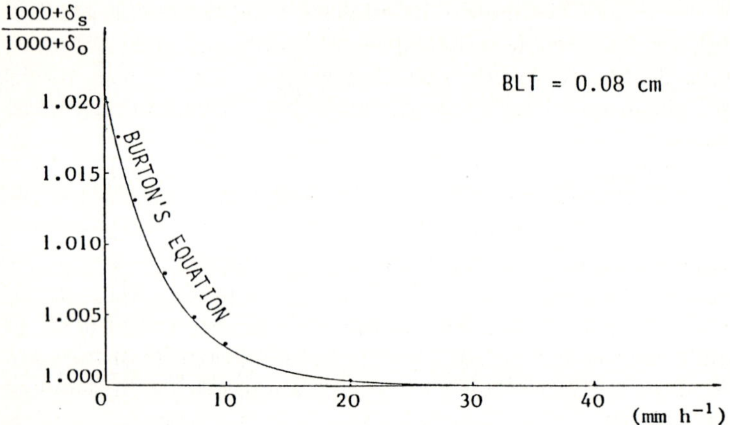

Fig. 2.22. Relationship between the freezing rate and the isotopic value of the steady state in the case of a semi-infinite reservoir. The *curve* represents Burton's equation, with a boundary layer thickness (BLT = 0.08 cm). δ_s and δ_0 are the isotopic values (in deuterium) in the ice and in the initial water respectively

$$\frac{1000+\delta_s}{1000+\delta_0}=\frac{\alpha}{\alpha-(\alpha-1)\ e^{-BLT\cdot V/D}}\ . \quad (8)$$

This equation is expressed graphically in Fig. 2.22 (for BLT = 0.08 cm) together with the points representing the steady-state values for the different freezing rates. Souchez et al. (1988a) have applied this to the determination of the growth rate of congelation ice in an Antarctic coastal area.

2.5 Impurities and Phase Equilibria

Ice contains impurities. Phase equilibria concepts are important to the understanding of their distribution.

Mulvaney et al. (1988) have shown that acids in the cold polar ice sheets may exist as aqueous mixtures at grain boundaries. With a scanning electron microscope equipped with a cold stage and an energy-dispersive X-ray microanalysis facility, they determined the location of sulphur in ice samples. As expected, sulphur was undetectable in the bulk of the ice; however, at the junctions where three grains are in contact, termed triple junctions, sulphur was found in concentrations of greater than 1 mole over areas of less than 1 μm^2. Such a concentration strongly suggested that the sulphur was present as a H_2SO_4/H_2O liquid mixture in equilibrium with ice at the temperatures of the ice sheet. That any liquid present would exist as veins at triple junctions rather than as films at two grain boundaries was predicted on the basis of thermodynamic considerations by Nye and Frank (1973). No chlorine was detected at the triple junctions examined, with the consequence that sodium chloride must have been distributed or localized within the grain. It is important to note that

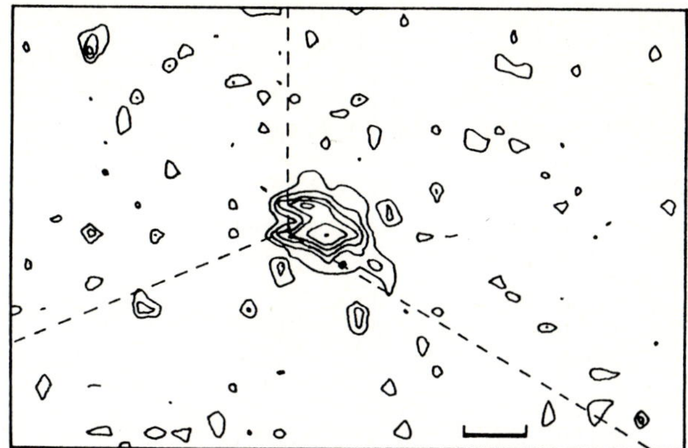

Fig. 2.23. The area around a triple-junction represented by the intersection of *three dashed lines.* The contour map represents the counts received at a X-ray detector for the channels corresponding to sulphur. Counts are registered at the spot where the electron beam entered the sample. The explored zone was gridded into squares $0.25 \times 0.25\ \mu m^2$. The mean and standard deviation (SD) of counts in each grid square was calculated. *Successive contour lines* represent the mean + n SD, where n has values of 1–6. *Bar* 1 μm. (After Fig. 3 in Mulvaney et al. 1988)

no signs of partial melting or recrystallization were observed in the ice analysed. Figure 2.23 from Mulvaney et al. (1988) shows the area viewed around a triple junction. It is a contour map of the counts received at the X-ray detector for the channels corresponding to sulphur. Successive contour lines represent the mean value + n standard deviations, where n is from 1 to 6.

For a better understanding of the distribution of impurities, let us compare the phase diagrams of a pure solvent like pure water and of a diluted solution with a molar fraction of component B, denoted x_B in Fig. 2.24. The vapour pressure curve of the solution is below that of pure water. This curve intersects the sublimation curve at a point defined by a lower temperature and a lower pressure than the triple point for pure water. Another value of the triple point is defined for the diluted solution. The sublimation curve for the solution corresponds to the one for pure water, since the two phases in equilibrium, solid and vapour, are identical in both cases; the vapour contains only pure water and, since the solution is diluted, freezing gives rise to pure ice. From the triple point corresponding to the dilute solution, a new freezing curve is displayed. This curve is practically parallel to the freezing curve of pure water, being slightly displaced towards lower temperatures. Under a pressure of one atmosphere, the freezing temperature will be lower.

If the two constituents (A and B) of a system can exist in the solid state over the temperature range considered, then the evolution of the system can be represented on a composition-temperature diagram, as in Fig. 2.25. Four distinct regions are displayed:

a) Below a well-defined temperature called the eutectic temperature, A and B coexist at equilibrium in the solid state.

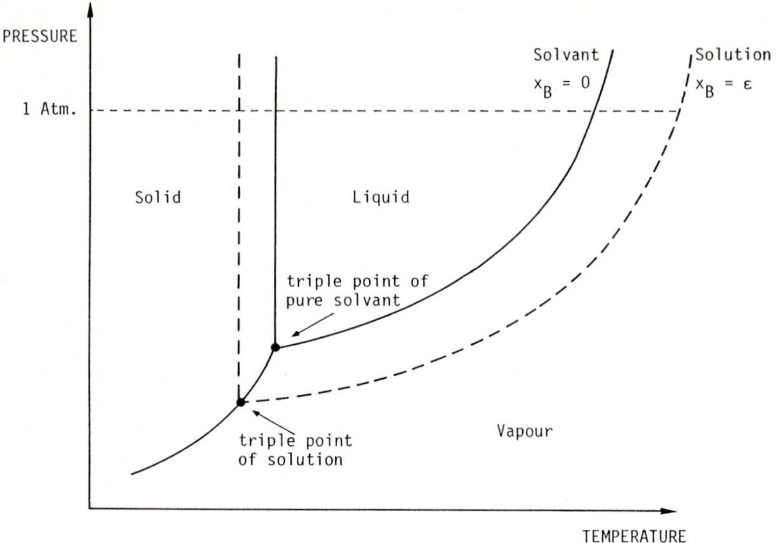

Fig. 2.24. Phase diagrams of a pure solvent and a dilute solution of component B. X_B is the molar fraction of component B

b) Above the eutectic temperature and above the two melting curves, the central region of the diagram represents a mixture of A and B, having melted and existing in the liquid form.
c) Above the eutectic temperature but below the melting curve of A, a lateral zone is displayed where A, in solid form, is in equilibrium with a mixture composed of melted A and B.
d) Above the eutectic temperature, but below the melting curve of B, a lateral zone is displayed where B, in solid form, is in equilibrium with a mixture composed of melted A and B.

Among these diverse regions, three represent domains where the phases coexist in equilibrium: either as a liquid and a solid or as a liquid and two solids.

If a mixture M composed in known proportions of solid A and solid B is progressively heated, the following evolution can be observed (Fig. 2.26):

- Up to the eutectic temperature, the two solids coexist, unchanged.
- At the eutectic temperature T1, a liquid phase appears. As the proportion of A in the initial mixture is greater than in the eutectic mixture, an excess of A will remain solid and the liquid will have the eutectic composition.
- At temperature T2, there is an equilibrium between the excess of solid A and a liquid of composition L. The proportions of liquid L and solid A are given respectively by the segments a and b. When the temperature increases, the liquid will be enriched in A and the quantity of solid will be reduced.

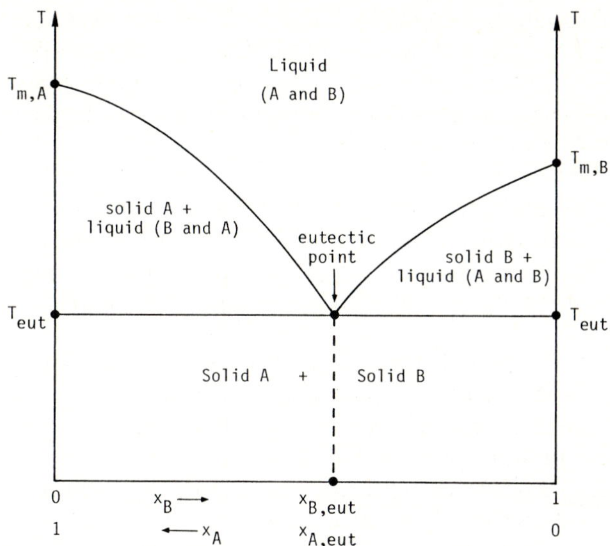

Fig. 2.25. Composition-temperature diagram for a solution of two constituents A and B. T_{eut} is the eutectic temperature. $T_{m,A}$ and $T_{m,B}$ are respectively the melting temperatures of the pure solid A and B. X_A and X_B are respectively the concentrations of constituents A and B at the eutectic point

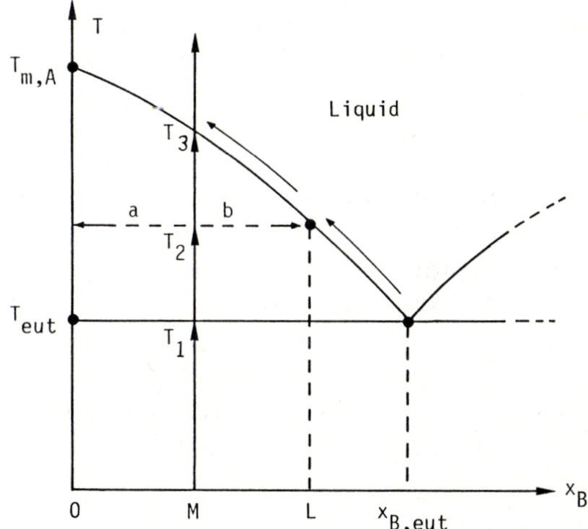

Fig. 2.26. Evolution of a mixture of a known proportion of solids A and B during heating from temperature T_1 to temperature T_3 (see text). $T_{m,A}$ is the melting temperature of the pure solid A

- At temperature T3, the composition of the liquid, having evolved along the melting curve, now has the initial composition of the solid and all the solid has melted at this stage.
- Above temperature T3, only the liquid phase is present.

In the temperature interval between T1 and T3, the solid and the liquid coexist and the melting of such a binary mixture is a progressive phenomenon, a situation completely different from that of a pure substance which melts at a given temperature. Only the eutectic mixture shows the latter characteristic.

If a melted mixture of A and B is progressively cooled the first appearance of a solid will produce a discontinuity in the rate of cooling because of the latent heat. Sometimes, the melting curves on a eutectic diagram show two distinct minima separated by a common maximum. This maximum always implies the formation of a stoichiometric compound of the two components A and B. This compound behaves like a new pure substance; it represents exactly the same situation as if two distinct eutectic diagrams were juxtaposed with a common temperature axis. Mixtures of sulphuric acid and water behave in this way.

The eutectic temperatures of different salts in aqueous solution are important to consider in the case of brines present between grains in glacier ice. Calcium carbonate is the first salt to precipitate at a temperature close to the freezing point of pure water followed by sodium sulphate in the form of mirabilite, at $-3.6\,°C$. The eutectic temperature of sodium chloride is about $-21\,°C$.

Selective rejection of solutes into the water by a growing solid phase is a general characteristic of most freezing systems. The effective distribution coefficient, defined as the ratio of the total concentration of solutes frozen into the ice to that remaining in the water, varies from 10^{-3} to unity. However, it is generally considerably less than unity (less than 10^{-2}) in water freezing at rates comparable to those expected within the glacial system. Thus, during freezing, nearly all the solutes present in the water tend to be preferentially rejected by the growing ice, which can be considered, as a first approximation, as a pure phase.

2.6 Self-Purification and Leaching of Impurities

Triple junctions, where water present as bulk liquid is in equilibrium with ice, form interconnecting networks in the ice. If, in a glacier at the pressure-melting point, water is moving through part of the vein system, as Shreve (1972) suggests, then impurities are removed from the ice. However, local movement of impurities and meltwater is unlikely to be able to purify the ice significantly. Shreve (1972) also indicates that large cavities and channels in the ice, as opposed to the small vein system, can be generated from the small veins as a result of pressure gradients in the grain-boundary liquid and frictional heating at the vein walls. Certain cavities continue to enlarge until they carry a substantial volume of water. These cavities intersect and form an arterial drainage system throughout the glacier. Water flow connected with such a drainage system does not give rise to a preferential loss of impurity content from the grains. The same is true for rapid flow through cracks, crevasses or moulins.

Glen et al. (1977) after reviewing the evidence for self-purification in temperate glaciers, suggest that the major purification mechanism occurs in liquid-like layers between grains and in the veins at triple junctions and that this takes place during recrystallization when ice is deformed. The ice-grain boundaries move through the material as the grain shape changes. The impurities reach the grain boundary by this process and it seems likely that such impurities will be retained within the boundary as it migrates. Freezing and melting accompany the recrystallization. As ice rejects most of the impurities to the meltwater when it freezes, the new ice that is formed will be purer. Finally, the impurities reach larger cavities and channels by which they are flushed out of the glacier. The question of whether vein flow is possible or is blocked by bubble formation or pinching off is not crucial. A vein system which allows free percolation will flush the impurities from the ice more quickly than one in which blocking occurs. On the other hand, Glen et al. (1977) expect grain boundaries to play a significant role, since a point in the bulk ice is surely more likely to meet a grain boundary than a vein and it is then more likely to travel with the boundary.

The flushing-out of cations from temperate ice is selective and fractional melting experiments (Souchez et al. 1973; Souchez and Tison 1981) have led to the recognition of this process. Slices from an Alpine glacier ice core, devoid of visible mineral particles, were heated from the base in a hot-water bath. As soon as 10 ml of meltwater was produced, it was recovered with the aid of a polyethylene syringe and immediately press-filtered. Successive 10-ml samples of filtered water were taken and analysed. Figure 2.27 shows a sharp decrease in the (Na + K)/(Ca + Mg) ratio during melting. This indicates a selective migration of ions, alkali metals being more easily flushed out than alkaline earth metals. Moreover, a systematic decrease in cation content is visible from the first to the last filtered sample. This is connected with dilution of interstitial water, which is mainly located at three- and four-grain intersections.

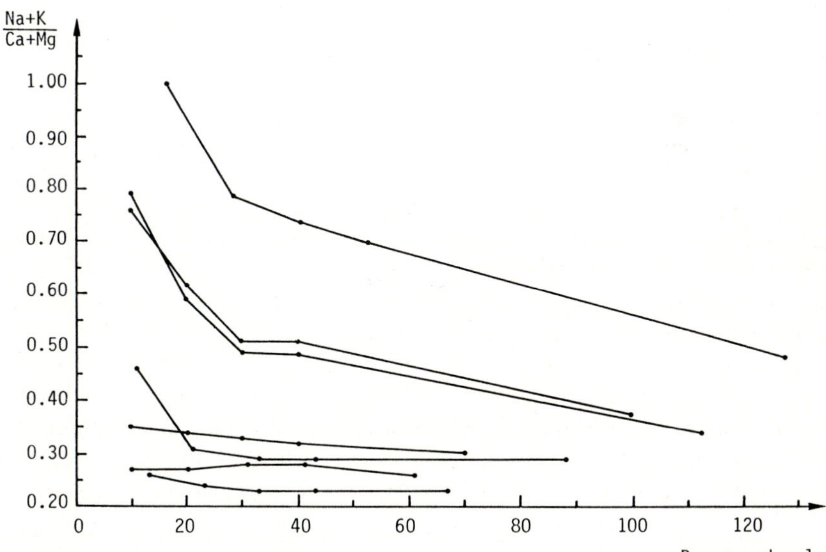

Fig. 2.27. The alkali-alkaline earth ratios for seven glacier ice samples subjected to fractional melting. (Souchez and Tison 1981, Fig. 3)

Free water, either rain- or meltwater from the surface layers, percolates through snow and firn. As it moves down into deeper layers, water-soluble substances originally present in the snow or firn as aerosols or dusts are entrained. This gives rise to a leaching process. In the French Alps, Ricq de Bouard (1977) has studied the distribution of major cations in a 6-m profile from a crevasse wall at 3200 m a.s.l., representing at least 20 years' accumulation. If the highly variable concentration of the first metre is not taken into account, Na, K, Ca and Mg appear to decrease exponentially with depth and the measurements indicate again that alkaline earth metals are less rapidly leached than the alkali metals.

Does leaching of impurities affect the distribution of radioactive products in ice in such a way to render them useless for dating purposes if meltwater is present? Let us develop the case of lead 210, which has been considered in studies on glacier dynamics. Lead 210 is produced in the atmosphere by the decay of ^{222}Rn. Air contains a certain amount of atmospheric aerosol-bound elements and lead 210 is brought to the ground principally by precipitation. The average concentration of ^{210}Pb in precipitation is of the order of five disintegrations per minute per litre. ^{210}Pb is a β-emitter, decaying to ^{210}Bi with a half-life of 22 years and ^{210}Po with a half-life of 5 days. Polonium 210 is an α-emitter, giving the stable ^{206}Pb with a half-life of 138 days. It is therefore

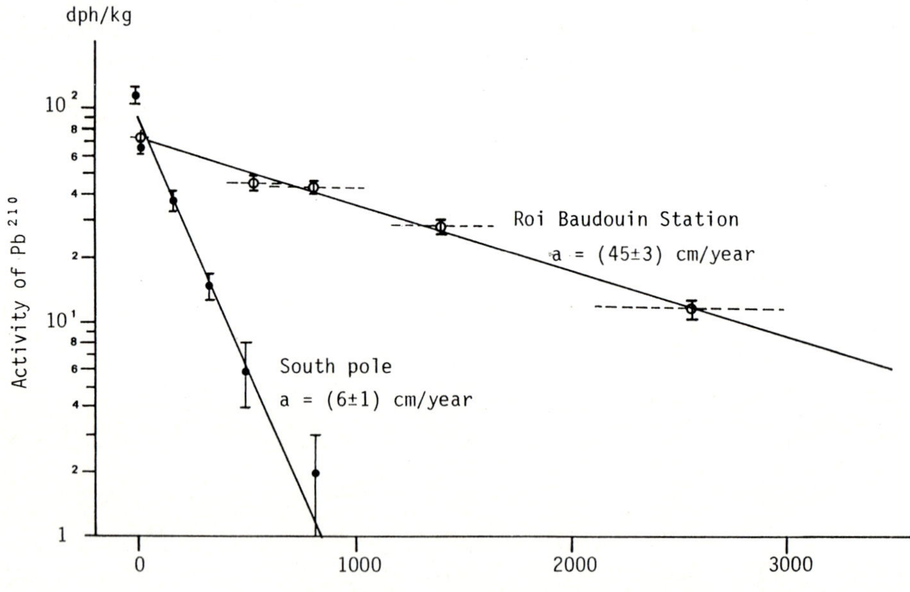

Fig. 2.28. Variation of the ^{210}Pb activity in the firn (disintegrations per hour per kilogram) as a function of equivalent water depth at South Pole and Roi Baudouin stations. The *vertical lines* indicate experimental errors in the ^{210}Pb measurements. The *horizontal dotted lines* indicate the thickness of the specimens. (Crozaz et al. 1964, Fig. 13)

possible to replace the measurement of ^{210}Pb by the measurement of the activity of one of its daughter products. For ice dating, the best method, requiring low-level counting equipment and elaborate radiochemical separation, is based on the α activity of ^{210}Po. The method relies upon the assumption that the specific activity of ^{210}Pb and the rate of accumulation remain constant. These conditions seem to be fulfilled in both polar ice sheets since, with proper sampling procedures, a linear decrease of the activity logarithm was observed at a number of stations when plotted against depth. Figure 2.28 from Crozaz et al. (1964) indicates this linear relationship both at the South Pole and at Roi Baudouin Station in Antarctica. Depth is expressed as equivalent water depth. The accuracy of the technique is in the order of ±10%. A similar graph has been obtained by Crozaz and Langway (1966) with ice samples from Camp Century in Greenland (Fig. 2.29). The ^{210}Pb dating method is valid in the absence of melting, with the result that each firn layer behaves as a closed system for ^{210}Pb, since the initial snow deposition. Nevertheless, Picciotto et al. (1967) have shown, at least for Kesselwandferner, a glacier in the Austrian

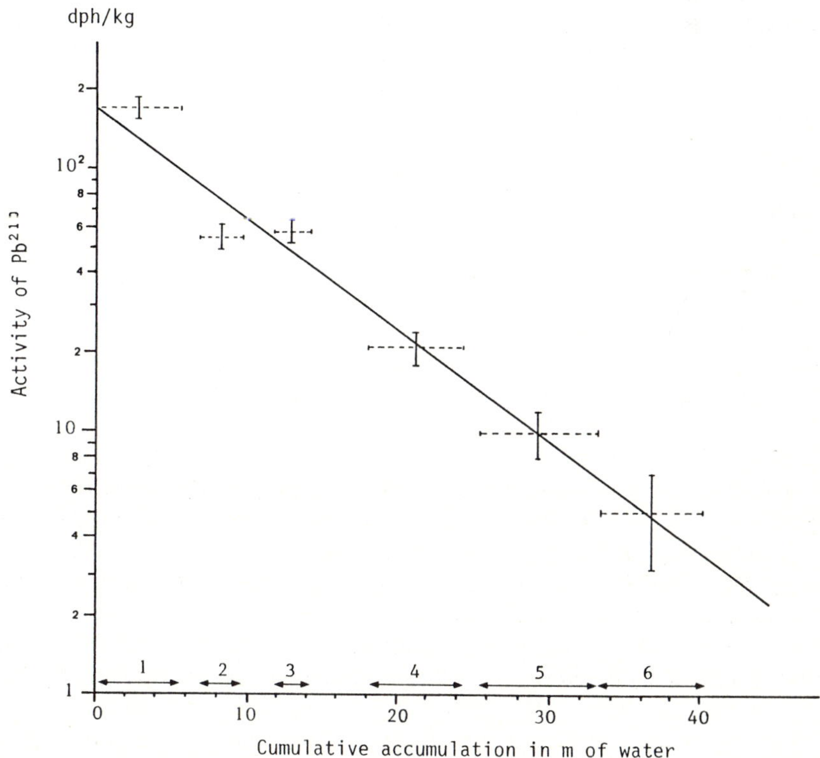

Fig. 2.29. Variation of the ^{210}Pb activity in the firn as a function of equivalent water depth at Camp Century, Greenland. The *vertical lines* indicate experimental errors in the ^{210}Pb measurements. The *horizontal dotted lines* indicate the thickness of the specimens. (Crozaz and Langway 1966)

Alps, that compact ice also behaves as a closed system. The lead 210 method is able to date ice up to 150 years old and, if the ice residence time is short, can give information on glacier dynamics. This will be developed in the chapter on ice composition and ice flow.

2.7 Mineral Particles in Ice

Material is added to the surface of glaciers and ice sheets from the atmosphere. Fallout of volcanic ash, deflation of sediment from dry, ice-free terrain, condensation nuclei in snowflakes and cosmic dust are examples of such contributions. Material is also added to glaciers and ice sheets by processes operating at the glacier-substrate interface. A distinction, common among glaciologists, exists between mineral particles of supraglacial and subglacial origin. Rockfall and avalanches from mountain sides and nunatak slopes above glaciers also contribute to the supraglacial debris load. If mineral particles are added in the accumulation zone, burial and englacial travel will ensue, while, if they are added in the ablation zone, mineral particles tend to remain at the surface, although debris falling into crevasses or carried into moulins by supraglacial streams can transport particles to deeper levels.

The 2164-m core drilled to bedrock at Byrd Station in Antarctica has revealed about 25 distinct tephra layers. The tephra layers contain glass fragments between 20 and 100 μm in size. The volcanic ash province of North Marie Byrd Land, a few hundred kilometers from the site, is the likely source. Small ice caps or valley glaciers also contain volcanic ash greater than 10 μm in size but none was detected in the Greenland Ice Sheet, although numerous high acidity layers are connected with Icelandic eruptions. A richer source of information for the present purposes is the microparticles ranging in size from 0.01 to 10 μm, contained within the ice. The examination of individual particles by scanning electron microscopy, coupled with elemental analysis by an X-ray energy-dispersive system, is a routine part of microparticle analytical procedure. These analyses are generally conducted for particles with diameters greater than 5 μm. Thompson and Mosley-Thompson (1981) have made a detailed study of microparticle concentration in the size range 0.6 – 0.8 μm in Dome C (East Antarctica), Byrd Station (West Antarctica) and Camp Century (Greenland) ice cores. Figure 2.30 gives the results respectively for five sections from the Holocene and for three sections from the Late Glacial portion of the Dome C ice core. Cyclical variations in insoluble microparticle concentration have been found to be annual features. However, problems that plague all stratigraphic interpretations, such as missing years, are present. On the basis of the data obtained, substantial increases in particle concentration are associated with more negative $\delta^{18}O$ values, representing the last major glaciation. Petit et al. (1981) have obtained a similar result from the study of the aluminium concentration in the ice together with its microparticle content. They have concluded that the glacial age climate was characterized by stronger atmospheric circulation, enhanced aridity and faster transport towards the Antarc-

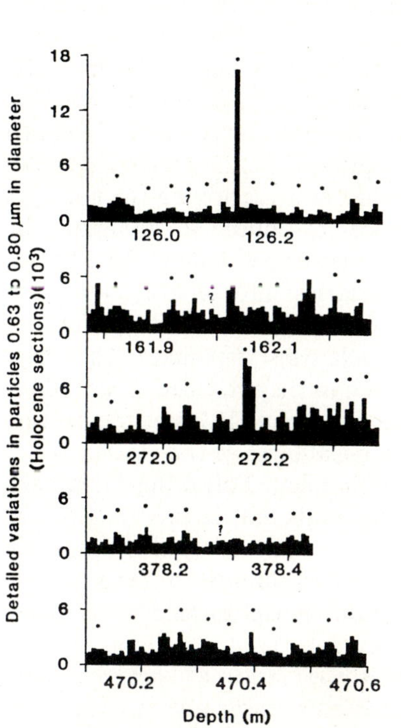

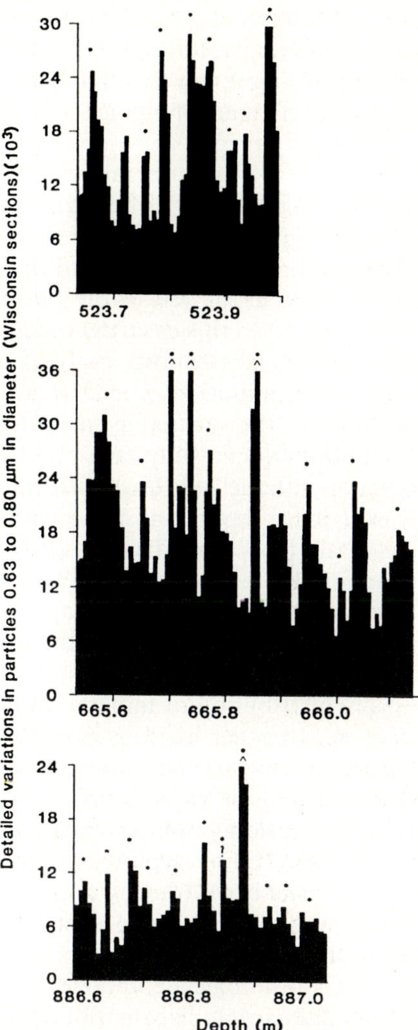

Fig. 2.30. Concentration of particles with diameters greater than or equal to 0.63 μm and less than 0.80 μm for five sections from the Holocene (upper 500 m) (*left*) and for three sections from the Late Glacial portion (lower 400 m) (*right*) of the Dome C ice core (Antarctica). *Dots* indicate annual features, *question marks* indicate questionable features, and ∧ indicates values exceeding the ordinate scale. (Thompson and Mosley-Thompson 1981, Fig. 1)

tic continent. Also to be considered is the exposure of continental shelf sediments by lowered sea levels. The problem will be analysed further when the question of the relative rheological properties of Holocene and Pleistocene ice is tackled.

Cosmic dust has been reported in Greenland ice since the 1950's, as summarized by Langway (1970). These particles are easily identified under a microscope, since they exhibit droplet-globules and spherical forms. As in the case of micrometeorites, their study is not crucial to the subject of this book but they do represent an atmospheric and therefore a supraglacial component of the debris load of the ice.

Mineral particles in basal ice are mostly of a subglacial origin. This is particularly clear if no nunatak or rock outcrop exists above the ice surface in the area investigated. Very often, grain size distributions of these particles are not strikingly different from those resting at the ice-substratum interface. This implies that they have been picked up by ice without modification of the properties acquired at the sole of the glacier and that they may eventually be recycled several times. In this case, the general principles governing the grain size distribution of basal tills also apply to the particles in basal ice.

The comminution process that occurs in the subglacial environment is dependent on the very large tractive forces which exist at the base of glaciers. Comminution can only take place if particles are either in contact with each other or with the bed. It is unlikely, for instance, that clasts dispersed englacially experience significant comminution, since the forces acting on the particles are small. Following Drewry (1986), two processes appear to dominate the comminution of clasts at the glacier bed : abrasion and crushing. Pure crushing causes disintegration along mineral boundaries into separate minerals, most mineral grains retaining their primary size during the process (Haldorsen 1981). Crushing thus produces a variety of particle sizes, depending essentially on size distribution of minerals in bedrock, often in the sand size mode. Abrasion, on the other hand, is defined by Sugden and John (1976) as the process whereby rock is scored by debris carried in the basal layer of the glacier. Abrasion can thus be considered as a wearing mechanism. Following Haldorsen (1981), abrasion produces cracks across the minerals and results in silt-sized rock flour. Milling experiments on sandstones carried out by this author in conditions either favouring crushing or abrasion lend support to this grain size differentiation by process: most of the sand-sized material formed by crushing, while the silt was mainly the result of abrasion.

Every lithological component of the particle population of basal ice shows a bimodal particle size distribution if the rock is monomineralic or consists of minerals with similar physical properties. One of the modes is in the clast-size group while the other reflects the mineral fragments in the matrix. The matrix mode can often be divided in two modes: one due to crushing and the other to abrasion. The ratio between the clast-size mode and the matrix mode is larger near the source, where the glacier has recently picked up the rock fragments. With increasing distance of glacial transport from the source, the matrix mode becomes more important and the clast-size mode is reduced. The matrix mode is restricted to certain particle size grades which are typical for each mineral. Terminal grades (Dreimanis and Vagners 1971) produced depend on the original size of the mineral grains in the rock and on the resistance of each mineral to comminution during glacial transport. Terminal grades for quartz range from 0.004 to 0.25 mm, for feldspars for 0.034 to 0.25 mm and for calcite from 0.002 to 0.062 mm. In this last case, solution during glacial transport must also be considered as well as the original particle size of the mineral and the effect of crushing or abrasion.

Some grain-size curves of particles finer than 3 mm in the basal ice of Alpine glaciers flowing on gneissic rocks exhibit a bimodal distribution. Figure 2.31a shows an example from Glacier de Tsijiore Nouve in the Swiss Alps. However, in the same environment, other size distributions are also present.

On phi-probability paper, a grain size distribution with a straight line will have a log-normal distribution. This type of distribution for clastic sediments results from a specific kind of fracturing process that is composed, as suggested by Epstein (1947), of a variety of breakage events. The model developed by Epstein leading to a log-normal distribution has two prerequisites:

a) The probability of fracture for each particle is a constant, independent of the size of the particle and of the number of breakage events that have occurred previously.
b) The size distribution obtained from the fracture of any single particle is independent of the initial size of the particle.

Very few grain-size distributions of particle populations in the basal ice of Alpine glaciers conform to such a log-normal distribution.

Kittleman (1964) has indicated that clastic materials derived through mechanical disintegration or crushing might be described better in relation to Rosin's law of crushing. Rosin's law was originally derived by Rosin and Rammler (1934) as a means of analysing artificially crushed material. This function is similar to the log-normal distribution in that it is strongly skewed but becomes more nearly symmetrical upon logarithmic transformation of the size variates. Rosin's law may be expressed as a straight line by twice taking the logarithm of the inverse of the cumulative weight percent and plotting the result as the ordinate and the logarithm of the size as the abcissa. Some grain-size distributions of particles from the basal part of glaciers in the Swiss Alps flowing on gneissic rocks obey Rosin's law of crushing (Fig. 2.31b as example). Dreimanis and Vagners (1971) suggest that, if several bimodal distribution curves are superimposed then the resulting curve becomes more or less straight on Rosin and Rammler paper. If, on the other hand, a monomineralic rock predominates, the cumulative curve does not approximate a straight line.

If the probability of fracture depends on the size of the particle, then crushing and abrasion will lead to a power function between cumulative weight percent and size. This power function can be transformed into a straight line on a bilogarithmic graph such that the prerequisite can be easily tested. The majority of the grain-size distributions examined by the authors in the area investigated in the western Swiss Alps on gneissic rocks exhibit a straight line on a bilogarithmic graph up to a few hundred μm (Fig. 2.31c as example). Thus, the role of size in the crushing and abrasion process seems important for the fine particles.

Studies on debris transport by glaciers make the distinction between two types of transport path: the "passive" one (supraglacial and englacial) and the "active" one (subglacial). Passive transport results in little modification of the

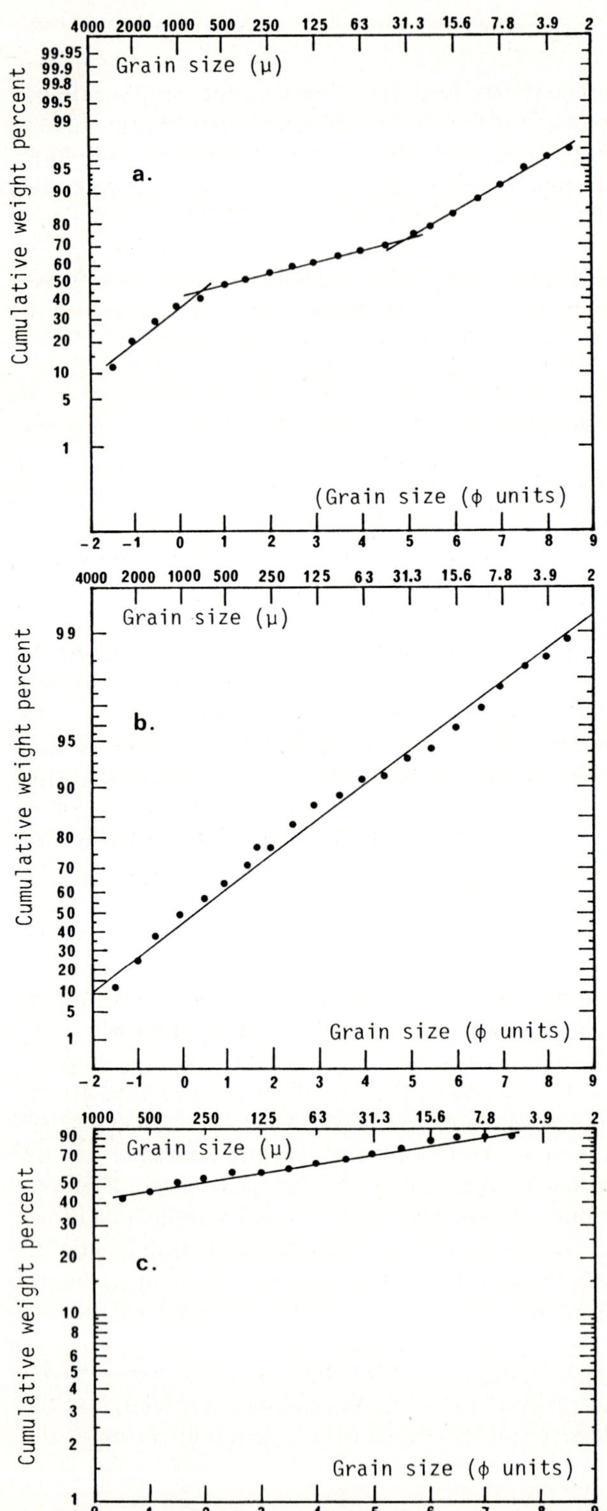

Fig. 2.31. Grain size distributions of particles embedded in basal ice from glaciers flowing on gneissic rocks in the Swiss Alps. **a** Bimodal distribution; ϕ-probability diagram. **b** Distribution obeying Rosin's law of crushing; Rosin-Rammler grid. **c** Bi-logarithmic distribution; bi-logarithmic graph

debris characteristics inherited from source material while active transport provides an environment in which new particles are produced and grains modified as a result of particle-particle and particle-bed contact. Clast shape is thereby modified, exhibiting striation, facetting and some degree of rounding (Reheis 1975; Drewry 1986). Concerning quartz sand grains, micromorphological studies have been carried out under the scanning electron microscope (Whalley and Krinsley 1974) to detect any features specific to the subglacial environment. However, Sharp and Gomez (1986) concluded that the examination of surface textures under the scanning electron microscope was unlikely to be a satisfactory technique for the discrimination between active and passive transport paths through glaciers. The relatively great hardness of quartz seems to protect it from significant modification by abrasive wear, the process most distinctive of the subglacial environment.

The petrography and mineralogy of debris in ice can also be used as tracers of processes of glacial erosion and transport and examples of this technique will be developed in the following chapters.

Besides, mineral particles in ice can influence its isotopic composition. Souchez et al. (1990) have studied a situation in which the isotopic characteristics of ice loaded with bedrock debris can be understood only if isotope exchange between hydroxyl-bearing minerals and water at the limit of ice grains is taken into account. Subglacial abrasion of micas and clays increases the number of broken bonds at the border of the mineral structure, promotes cleavage and so increases the rate at which these minerals exchange oxygen and hydrogen isotopes with water. The $\delta^{18}O$ values of hydroxyl-bearing minerals range between +7 and +27‰ with clays more positive than micas; the δD values for both minerals range between −40 and −100‰. If the water volume relative to the amount of hydroxyl-bearing minerals is small, the isotopic composition of the water is likely to be markedly affected by a process of isotope

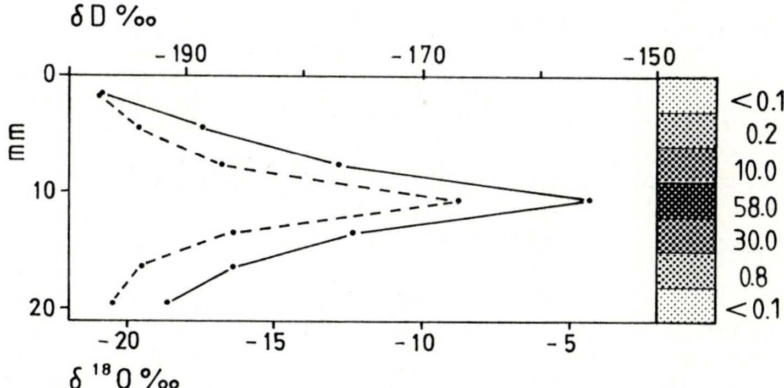

Fig. 2.32. Isotopic composition and debris/ice ratios in a thin zone of a debris-rich ice layer at Qigssertâq, Greenland. The *solid line* refers to the δD scale, the *dashed line* to the $\delta^{18}O$ scale. The measured debris/ice ratios are given (wt ‰) on the *right*. *Shading* is a visual representation of the debris concentration. (Souchez et al. 1990, Fig. 3)

exchange. For example, if ground material comes into contact with a limited amount of meltwater from Pleistocene ice, i.e. ice with very negative δ values, the water will become enriched in heavy isotopes. A 2-cm-thick debris-loaded ice layer from the basal zone of the Greenland Ice Sheet in Qigssertâq (Fig. 1.15b) was investigated in detail with a microtome in a cold room. Figure 2.32 shows results of the analyses. Considerable enrichment in heavy isotopes occurs when the debris/ice ratio is particularly high. The influence of this process has thus to be considered when studying the isotopic composition of basal ice.

2.8 Gases in Ice

Practically all gas in natural ice is concentrated in air bubbles so long as the cryostatic pressure does not exceed 80 bars. Bubble free ice contains less than 0.0001 cm^3 gas (at standard temperature and pressure) per gram of ice, i.e. 10^{-3} times the concentration usually found in cold bubbly ice.

When firn is transformed into ice near the surface of an ice sheet, air is sealed inside the ice. The process occurs continuously with time and the trapped air moves as the ice moves. The age of the entrapped air is, however, not the same as that of the surrounding ice because air bubbles only become isolated from the atmosphere during the transition from firn to ice. Following Schwander and Stauffer (1984), the age of the ice at this transition is typically between 100 and 3000 years. Table 2.2 gives the mean age difference between ice and enclosed air at a number of polar stations. This parameter increases as the mean annual air temperature decreases. Any initial gas content in the

Table 2.2. Measured values for Siple Station, and calculated values for other important sites, of the difference between the age of the ice and the mean age of the air trapped in its bubbles as well as the age distribution width. (Schwander and Stauffer 1984, Table 1)

Site	Location		Mean annual air temperature (°C)	Annual accumulation (m water equivalent)	Width of age distribution (a)	Mean age difference between ice and enclosed air (a)
					Measured	
Siple	75°55′S,	83°55′W	−24	0.5	22	95
					Calculated	
Vostok	78°28′S,	106°48′E	−57	0.022	590	2800
Dome C	74°39′S,	124°10′E	−53	0.036	370	1700
Southpole	90°00′S		−51	0.084	220	950
Byrd Station	79°59′S,	120°01′W	−28	0.16	54	240
North Central	74°37′N,	39°36′W	−31.7	0.11	76	350
Crête	71°07′N,	37°19′W	−30	0.265	46	200
Camp Century	77°11′N,	61°09′W	−24	0.34	31	130
Dye 3	65°11′N,	43°50′W	−19.6	0.5	22	90

ice grains is small compared to the value of about 0.1 cm^3/g of ice generally present as bubbles in polar glacier ice. Thus, the composition of gas bubbles in cold ice from polar glaciers is close to atmospheric. Gas content and composition analysis can therefore provide information on the environmental conditions at the time of formation of the glacier ice (Stauffer et al. 1984; Stauffer et al. 1985 a; Barnola et al. 1987).

By contrast, ice formed in the presence of meltwater is different. Considerable variations in gas content and composition have been observed and the gas composition can deviate significantly from the composition of the atmosphere. Water in equilibrium with the atmosphere dissolves 0.03 cm^3 of air/g at standard temperature and pressure. If meltwater saturated with air penetrates into cold snow, it refreezes as a thin layer of ice covering cold snow grains or, only in the case of heavy meltwater production, as a superimposed ice layer. The number and size of the bubbles depends mainly on the freezing rate. The gas composition can be close to the composition of air dissolved in water which is different from atmospheric air: ratios between selected gases reflect their relative solubilities, the more soluble being preferentially enriched. Therefore CO_2 is the most enriched, followed by argon, then oxygen and finally nitrogen. However, the gas composition in the bubbles depends additionally on the diffusion constants of the different gas components in water since, during refreezing, part of the dissolved gases escape by diffusion (Stauffer et al. 1985 b). Compared with Ar, N_2 or CO_2, the diffusion of O_2 is much faster (Houghton et al. 1962). These processes may take place in the firn where the pore space between the crystal grains is filled with air and water.

Water within temperate glacier ice originates partly from the surface and partly from internal melting due to frictional heat. Surface water is normally saturated with air and probably has little effect on the gases in the bubbles. Heat of internal friction produces practically gas-free meltwater. Upon making contact with gas bubbles, this meltwater will dissolve part of the air and, due to the intergranular drainage system examined in Section 2.6, removal of gas from temperate glacier ice occurs in this way. Even if the water is saturated at a certain depth, it is still able to dissolve more gas at greater depths since the gas pressure in the bubbles is higher. The result of this removal process during glacier flow is a depletion of the more soluble gases. Weiss et al. (1972) indicate that samples from the terminus of the Aletsch Gletcher in the Swiss Alps are depleted in total gas content by a factor of twenty compared with samples from the source region and that Ar and O_2 have been preferentially removed compared to N_2. A consequence of the process is the transformation of bubbly ice into less bubbly and eventually clear ice along the flow line.

Measurements of the total gas content in polar ice can give information on the elevation at which the ice was formed, as developed at the beginning of the next chapter. This is possible because differential migration of air does not occur in cold glacier ice and also because the volume of the pores can be known for the time they close off from the atmosphere at the ice formation site. Developments of this method will be given in the next chapter owing to its potential importance in studies of ice flow.

Part II Implications on Glacier Dynamics

3 Ice Composition and Ice Flow: a General View

3.1 Gas Content and Ice Sheet Profiles

From the perfect gas law, one can write

$$V \cdot \frac{P}{T} = V_0 \cdot \frac{P_0}{T_0} , \quad (1)$$

where V is the total amount of gas present per gram of a polar ice sample at the standard temperature T and the standard pressure P; V_0 is the pore volume when the pores are finally closed off in the firn at temperature T_0 and air pressure P_0. The temperatures are expressed in K (Raynaud and Lebel 1979). This is only true if the transformation of snow into ice occurs in the absence of meltwater, i.e. in the dry snow zones of polar ice caps and ice sheets.

T_0 is the temperature of the ice just formed; it depends on surface temperature. The depth at which firn is transformed to ice in a polar glacier varies from 50 m near the coast to 150 m in the central regions of Antarctica. At 10 m depth, seasonal temperature variations are reduced to 1% of surface temperature variations. Below 10 m, temperature gradients are between 0.01 °C per metre near the coast and 0.02 °C per metre in the central regions (Raynaud 1976). Thus the difference between the close-off temperature T_0 and the 10-m temperature is about 3 °C as maximum. In turn, the 10-m temperature is strongly correlated to elevation as indicated by measurements taken along a flow line in Terre Adélie, Antarctica (Fig. 3.1).

P_0 is the pressure in the firn pores when the air becomes occluded in the ice and is a function of the surface atmospheric pressure. Variations of surface pressure will be more or less damped at depth. Monthly mean pressure fluctuations at a site are within 4% of the mean annual value. Table 3.1 from Raynaud (1976) gives these values for a number of Antarctic stations. The mean pressure in the coastal zone of Antarctica is 980–990 mb while it falls to 600 mb at 4000 m elevation with a vertical gradient between 9.5 and 13 mb per 100 m. Thus, like T_0, P_0 is mainly dependent on the elevation parameter.

V_0 is the gas volume of the ice just formed. It is considered by Raynaud and Lebel (1979) to be dependent simply on T_0; the empirical relationship found by these authors being

$$V_0 = 7.4 \cdot 10^{-4} T_0 - 0.057 ,$$

where V_0 is expressed in cm^3/g of ice and T_0 in K.

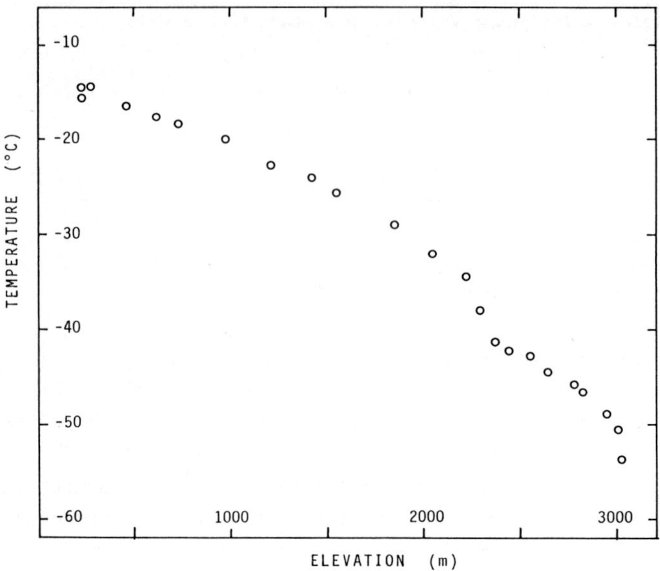

Fig. 3.1. Temperature of the firn at 10 m depth as a function of elevation along a flow line in Terre Adélie (Antarctica). (Raynaud 1976, Fig. 5)

Table 3.1. Climatic data from different Antarctic stations. (Raynaud 1976, Table 3; data from Schwerdtfeger 1970)

Station	Elevation (m)	T Mean annual temperature (°C)	$P \pm \sigma$ Mean annual atmospheric pressure σ standard deviation[a]	ΔP (mb) Maximum range of mean monthly pressure	$\Delta P/P$ (%)	P(mb)/T(K)
Plateau 79°15′S – 40°30′E	3625	−56.4	609 ± 7	22	3.6	2.810
Vostok 78°28′S – 106°48′E	3488	−55.6	624 ± 5.5	15.5	2.5	2.868
South Pole 90°00′S	2800	−49.3	681 ± 5	13	1.9	3.042
Pionierskaya 69°44′S – 95°30′E	2740	−38.0	690 ± 6.5	17	2.5	2.934
Byrd 80°S – 120°W	1530	−27.9	806 ± 6	17	2.1	3.286
Eights 75°14′S – 77°10′W	421	−26.0	942.5 ± 4.5	12	1.3	3.813
General Belgrano 77°58′S – 38°48′W	50	−22.3	986 ± 3.5	12	1.2	3.930
Little America 78°18′S – 163°00′W	40	−20.9	982 ± 6	18	1.8	3.893

[a] Standard deviations for P values are calculated from mean monthly values.

Substitution of this equation into Eq. (1) gives:

$$V = \left[7.4 \cdot 10^{-4} P_0 - 0.057 \frac{P_0}{T_0}\right] \cdot \frac{T}{P} .$$

Thus, the total amount of gas trapped in polar glaciers, as dry firn is transformed into ice, depends on the prevailing atmospheric pressure and temperature at the formation site. In present-day surface conditions, an elevation increase of 1000 m corresponds to a decrease in V of about 0.017 cm^3/g of ice while a general cooling of 5 °C at the formation site would imply a decrease in V of approximately 0.001 cm^3/g of ice. Thus, the total gas content of ice V is a good indicator of the elevation of the site of formation of the ice considered. So, gas content determinations can provide information relating to sites of ice formation and, by extension, to past variations in the thickness of polar ice sheets.

One of the important tasks faced by ice flow studies is that of reconstructing ice sheet profiles at the time of the Last Glacial Maximum, about 18,000 years ago. For the Byrd ice core, Raynaud and Whillans (1982) calculated the expected V profile from steady-state conditions, assuming a surface elevation profile constant with time, upstream from Byrd Station. The results are given in Fig. 3.2. The V measurements from the Byrd core are plotted on the same figure. A significant deviation exists between the steady-state curve and the V measurements in the depth interval corresponding to ice from the Last Glacial Maximum as indicated by the δ profile. This deviation suggests that the elevation of the site of formation of the ice, estimated to be about 100 km upstream from the drilling site, was lower than present by about 200 m (Lorius et al. 1984). These conditions may be typical for the central area of West Antarctica. A similar study carried out on the Vostok ice core in the central part of East Antarctica gives an elevation about 100 m lower than that of the present day. This indicates a slight thickening of this central part of East Antarctica since the Last Glacial Maximum but, in any case, no dramatic change. This slight thickening, along with that of the central part of the West Antarctic Ice Sheet between the Last Glacial Maximum and the present, is probably connected with an increase in the rate of snow accumulation. The interpretation of combined V-δ profiles, like the one shown in Fig. 3.2, is more difficult in coastal Antarctic sites because of ice-flow effects. Nevertheless, results from Terre Adélie and from the Law Dome near Casey (66°17′S, 110°32′E) suggest that the ice thickness was in the order of 400 m greater during the Last Glacial Maximum than at present. Coastal areas of the East Antarctic Ice Sheet therefore seem to have thinned between the Last Glacial Maximum and the present. This may be due to the influence of the sea-level rise and to the temperature increase.

For the Greenland Ice Sheet, a similar study has been conducted on the Camp Century ice core (Raynaud and Whillans 1979). Figure 3.3 shows the $\delta^{18}O$ and the V profiles for this core. The principal feature of the record of elevation change with time is an important decrease of about 900 m between

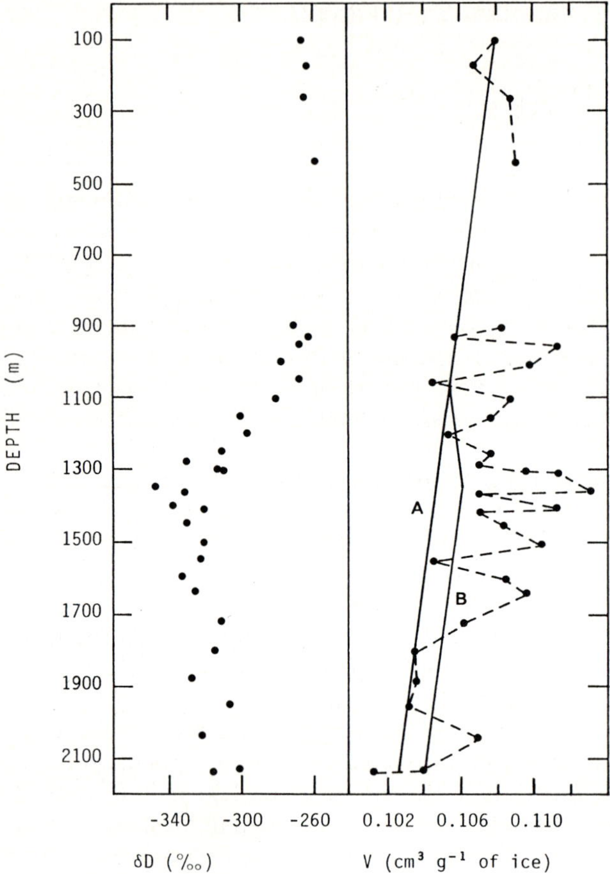

Fig. 3.2. Variations of the mean total gas content V and δD with depth at Byrd station. Mean V values measured at different depth levels are compared with the change of V to be expected by ice flow according to the present flow pattern under constant temperature (line A) and corrected for colder temperatures during the ice age (curve B). (Raynaud and Whillans 1982, Fig. 3)

10,000 and 5000 years b.p. During this time interval of 5000 years, a change of only about 60 m can be attributed to ice flow according to the current flow pattern. Consequently, there must have been a substantial change in the ice sheet geometry over this period. A change of accumulation rate cannot explain the situation: a 50% decrease in the accumulation rate would account for a thinning of only 200–300 m. The authors favour a model involving the rise of sea level. The melting of the Laurentide Ice Sheet in North America and of the Scandinavian Ice Sheet in Northern Europe has produced a sea level rise that could have affected the marginal part of the Northwest Greenland Ice Sheet. The margin of the ice in the area of Camp Century is at sea level in Melville Bay today. The simplest reconstruction of the ice sheet during the Last Glacial Maximum considers a similar shape as today and allows a marginal ex-

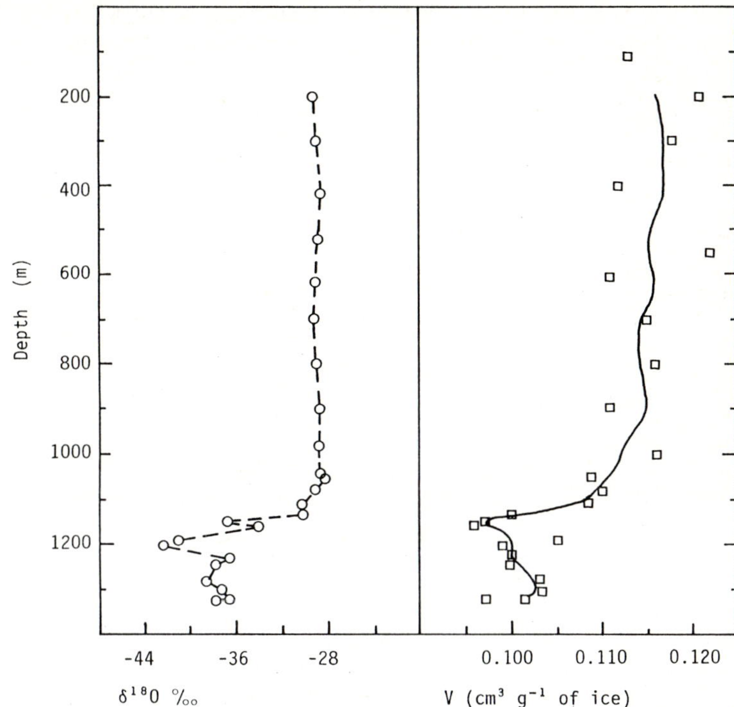

Fig. 3.3. Variations of the mean total gas content V and $\delta^{18}O$ with depth at Camp Century. *Open squares* give mean values of V for each level analysed. The *solid curve* gives the smoothed variation. (Raynaud 1976, Fig. 8)

tent in Melville Bay corresponding to a 120–130-m lower sea level. Raynaud and Whillans (1979) calculated that such a change can account quantitatively for the elevation changes deduced from the V measurements. The study of the total gas content of ice which has never been at the melting point during or after its formation can thus be an effective tool in reconstructing ice sheet geometries through time.

3.2 Impurities in Ice and Ice Creep

There is little doubt that the existence of impurities in ice will influence its creep rate. Jones and Glen (1969) have studied the effects of dissolved impurities on the mechanical properties of ice crystals. The addition of various chemicals to crystals of ice – a process called doping – has, in general, the effect of "softening" the ice and enhancing the creep rate. This is probably due to the fact that the eutectic temperatures of a great number of soluble salts are very low so that salt-rich solutions exist between the ice crystals and increase sliding at grain boundaries.

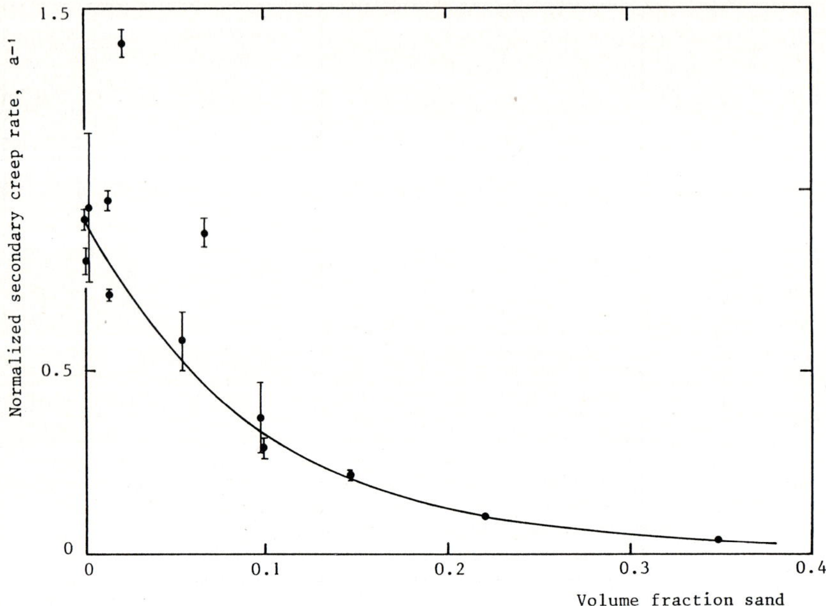

Fig. 3.4. Creep rate as a function of the particle content of ice: results of experiments. *Error bars* represent 99% confidence limits on slope of secondary creep curve. *Error bars on the two samples in the lower right corner* are smaller than size of dot. Points for clean ice offset horizontally for clarity. (Hooke et al. 1972, Fig. 3)

Concerning solid impurities in ice, sand in concentrations about 0.5 – 15% by volume will lead to a significant increase in strength ("hardening"), possibly in connection with a decrease in crystal size. Particle impurities at grain boundaries block the movement of dislocations and also inhibit grain growth, thereby reducing the creep rate. Hooke et al. (1972) studied the creep of ice containing dispersed fine sand at a temperature of about −9 °C. Results are given in Fig. 3.4. Over the range considered, the creep rate decreased with an increase in sand concentration. However, field data of Echelmeyer and Zhongxiang (1987) indicate debris-rich ice is softer than clear ice and boudinage structures in basal ice also show that dirty ice is softer.

Wisconsin-aged ice is mechanically softer than Holocene ice, at least in the Northern Hemisphere. Borehole deformation studies confirmed this rheological contrast on three main ice bodies: the Devon Ice Cap (Paterson 1977), the Agassiz Ice Cap (Fisher and Koerner 1986) and the Greenland Ice Sheet at Dye 3 (Dahl-Jensen and Gundestrup 1987). On the Barnes Ice Cap, Hooke (1973 a) carried out borehole deformation experiments and showed that shear deformation rates in a layer of bubbly white ice at the base of the ice cap were about 2.5 times greater than in the overlying blue ice. Subsequently, isotopic data indicated that the boundary between the white and blue ice zones represented the Pleistocene-Holocene transition (Hooke 1976; Hooke and

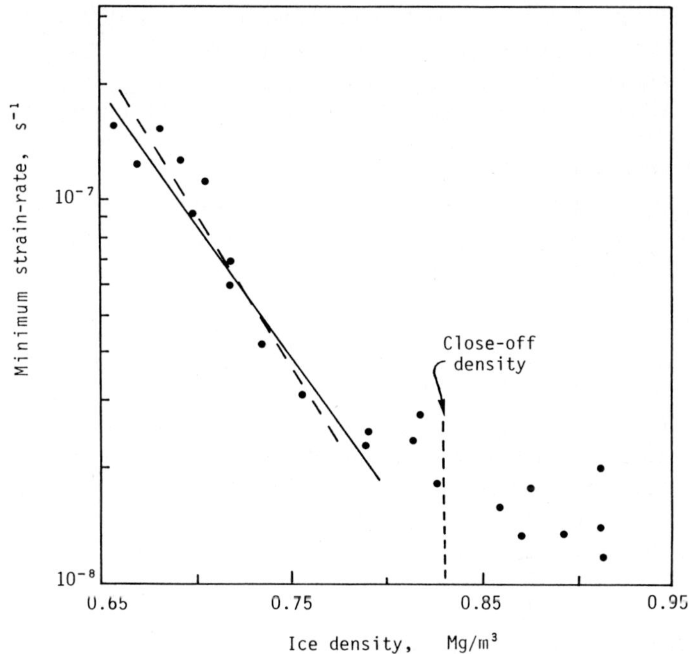

Fig. 3.5. Variation in minimum shear strain-rate with initial ice density. *Slopes of solid and dashed lines* are based on tests of Mellor and Smith (1967) and Haefeli and Von Sury (1975) respectively. (Hooke et al. 1988, Fig. 1)

Clausen 1982). Hooke (1973a) has attributed the softness of the white ice to its high bubble content which results in a density difference of 0.05 g/cm^3. Based on extrapolation of creep data for low-density ice, Hooke's opinion was that this density difference was quantitatively sufficient to explain the contrast in rheology. Recent experimental studies by Gao and Jacka (quoted in Hooke et al. 1988) indicate no significant change in minimum strain-rate over a density range from 0.83 to 0.91 g/cm^3. Figure 3.5 summarizes the results obtained. This suggests that the rheological contrast between Pleistocene and Holocene ice in the Barnes Ice Cap should be attributed to factors other than the density contrast. The change in fabric and crystal size, which is dramatic across the Pleistocene-Holocene boundary, in addition to the chemical contrasts in the ice cap (Table 3.2) are other factors that must be taken into account in order to explain this rheological contrast.

Table 3.2. Summary of chemical analyses of Barnes Ice Cap ice, ppm. (Hooke et al. 1988, Table 1)

	Na	K	Ca	Mg	Cl
Holocene ice	0.024 ± 0.010	0.012 ± 0.007	0.010 ± 0.015	0.002 ± 0.001	<0.05
Pleistocene ice	0.049 ± 0.043	0.043 ± 0.045	0.244 ± 0.231	0.044 ± 0.043	0.1 ± 0.1

Ice from the Last Glacial Maximum is usually much richer in impurities than Holocene ice. For example, at Dome C in East Antarctica, it contains about 1000 ng of impurities per gram of ice, 50% of them being soluble, whereas Holocene ice only contains 200 ng of impurities per gram of ice, 90% of which are soluble. Thus, the difference concerns soluble salts as well as insoluble microparticles. The influence of the contrast in impurity content of ice of different ages on ice flow is only now emerging and, no doubt, further studies will shed more light on this problem and its influence on the behaviour of ice sheets and ice caps.

3.3 Isotopes and Flow in Ice Sheets and Ice Caps

Until recently, little attention has been paid to isotopic measurements in the ablation zone of ice sheets, in spite of the fact that this area is readily accessible in comparison with the central region.

Figure 3.6 gives idealized flow lines from a steady-state theoretical ice sheet. The main features are the marginal position of the ablation area, the downward movement of ice in the accumulation area and the upward movement in the ablation area. The vertical velocity component is directed downwards in the accumulation area. As a consequence of this vertical movement and outward displacement, annual layers of ice are thinned as one goes from the surface of the ice sheet to the basal zone. In the ablation zone, where ice is removed from the surface as a result of melting or sublimation, the vertical velocity component is directed upwards, compensating for this loss in the steady-state profile. At the centre of the ice sheet, ice is being added to the top surface at the accu-

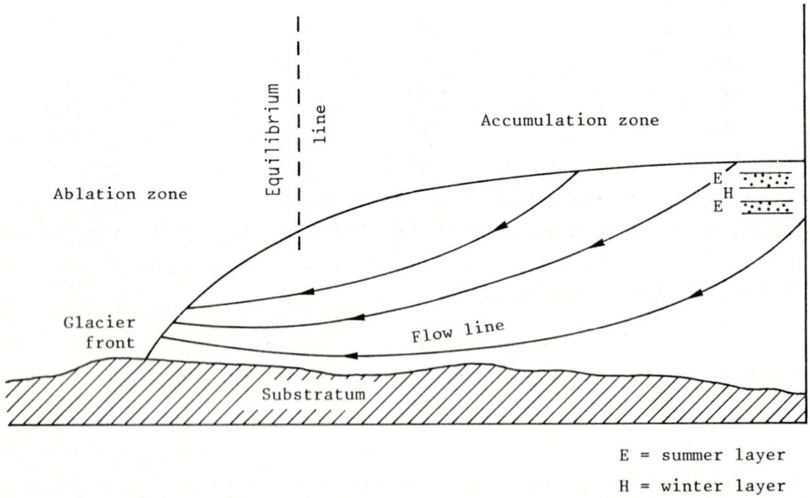

Fig. 3.6. Idealized flow lines in a steady-state theoretical ice sheet

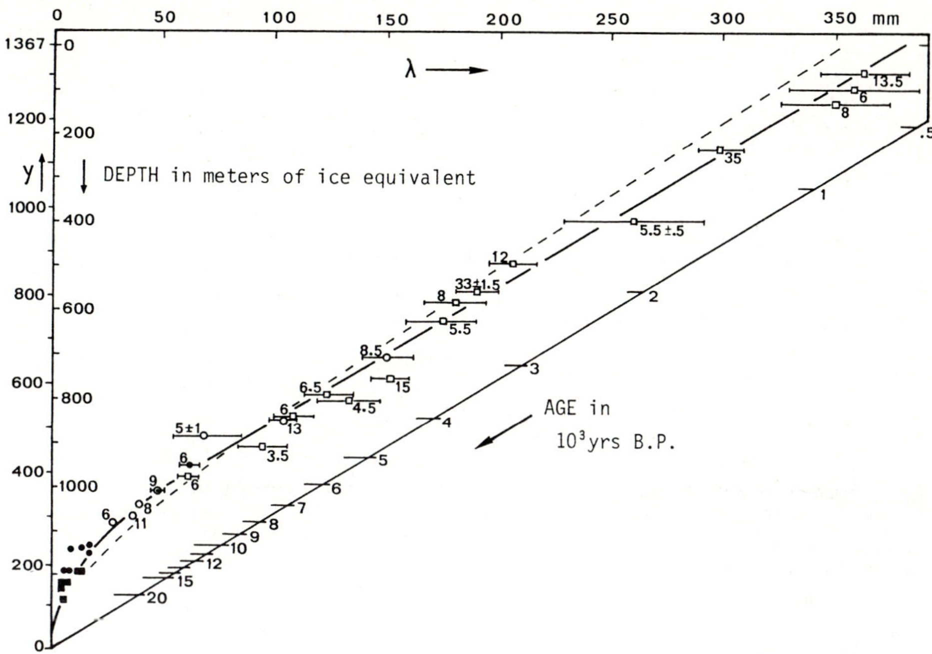

Fig. 3.7. Annual layer thickness, λ, in the Camp Century ice core plotted against depth and distance y above the bottom. The *heavy curve* is the least-squares fit to all data with y >250 m. The corresponding time scale is plotted along the sloping axis. The *dashed curve* is the least-squares fit, when using Philberth and Federer's (1971) procedure. The *figures close to the points* are the number of annual layers interpreted in the individual core increments, either directly from $\delta(^{18}O)$ profiles (*open squares*), or after deconvolution (*open circles*), or from measured microparticle profiles (*filled circles*), or just by visual stratigraphy (*filled squares*). (Hammer et al. 1978)

mulation rate. The surface must be sinking at the same rate for the thickness to remain constant. To compensate for this downward motion, the ice column must spread out horizontally at a uniform rate at all depths. This corresponds to a vertical velocity that falls uniformly from a value dependent on the accumulation rate at the surface to zero at bedrock at the centre of the ice sheet. Away from the centre, the decrease of the annual layer thickness will also occur but not linearly. As indicated in Fig. 3.7 from Hammer et al. (1978), annual layer thickness in the Camp Century ice core in Northwest Greenland fits with this simple model, assuming a linear variation with depth to about 400 m above the bedrock, corresponding to a uniform vertical strain rate. Below this level the strain rate decreases linearly to zero at bedrock, the interface being largely below the pressure-melting point.

Raynaud (1976) presents isotopic measurements from the western margin of the Greenland Ice Sheet, along the E.G.I.G. line. The E.G.I.G. line is a geophysical traverse made from the surroundings of Camp VI on the ice sheet, north-east from the city of Jakobshavn to the ice divide at Crête. Intermediate stations are Milcent and Station Centrale (Fig. 3.8). The δD and $\delta^{18}O$ mea-

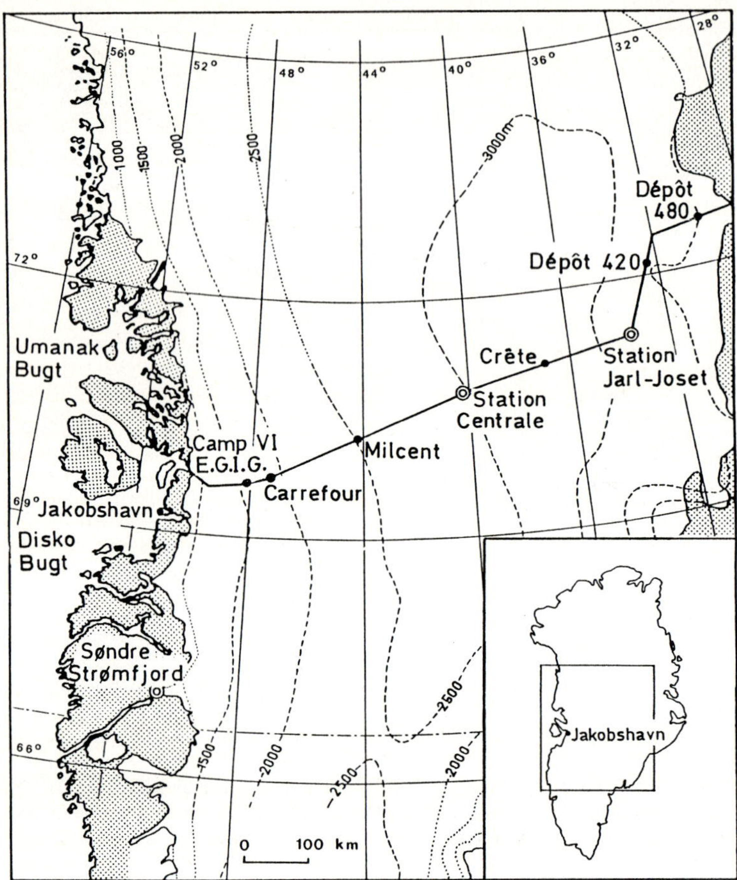

Fig. 3.8. Map of Greenland showing the E.G.I.G. line. (After Fig. 1 in Merlivat et al. 1973)

surements are plotted on Figs. 3.9 and 3.10 versus the distance from the moraine located at the western edge of the ice sheet.

Figure 3.9 gives the variation in isotopic content from the moraine to a point located 500 m upstream in the outer ablation zone. Figure 3.10 shows the variation in isotopic content from 0 to 50 km over the ablation zone (Fig. 3.10a) and shows variations from 0 to 500 km on a different distance scale but with the same δ value scale (Fig. 3.10b). Values inland of the equilibrium line have been measured at the sites indicated. Horizontal lines link locations with same observed surface mean δ values in the accumulation and ablation zones. By linking sites in the ablation zone with the corresponding points on the inland E. G. I. G. profile at which surface firn has the same δ value, an estimation of the steady-state model can be checked. In a two-dimensional model, let dx_A and dx_N be a section of the accumulation zone and a section of the ablation zone respectively, separated by a mean distance x_A and x_N

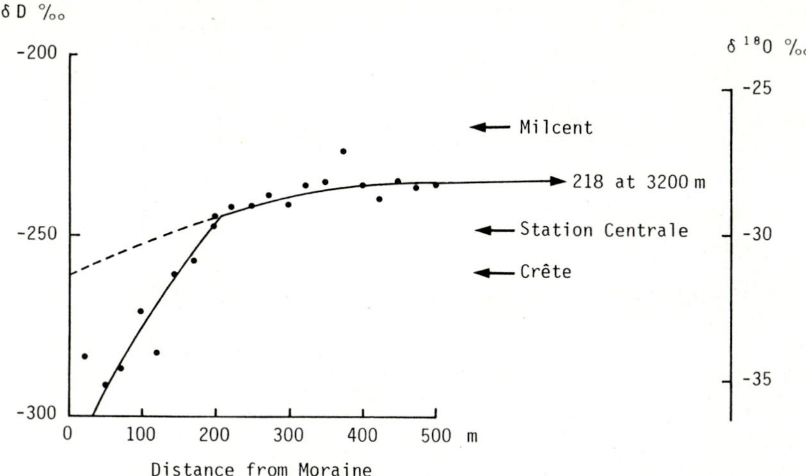

Fig 3.9. Variation of δD values with distance inland from the moraine at the western edge of the Greenland ice sheet at 69°43′N. Each *point* is the mean of corresponding values along two parallel sampling lines 100 m apart. The *right-hand scale* shows corresponding δ^{18}O values. (Data from Raynaud 1976). (Robin 1983b, Fig. 4.13)

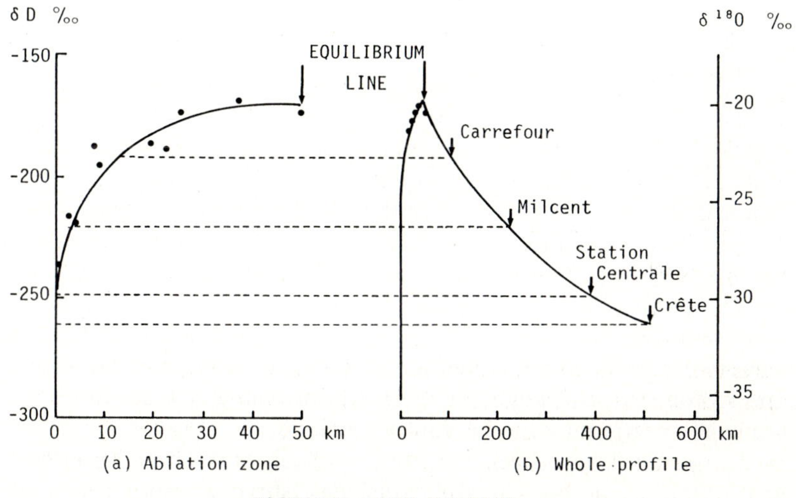

Fig. 3.10. Variation of isotopic δD values with distance from the edge of the Greenland ice sheet (see text). (Data from Raynaud 1976, Merlivat et al. 1973). (Robin 1983b, Fig. 4.14)

from the equilibrium line (Fig. 3.11). A and N are the accumulation rate and the ablation rate in dx_A and dx_N respectively. Then, following Robin (1983a), for the surface of the ice sheet to be in a steady state, the ablation rate over dx_N must equal the accumulation rate over dx_A:

$$A \cdot dx_A = N \cdot dx_N \ .$$

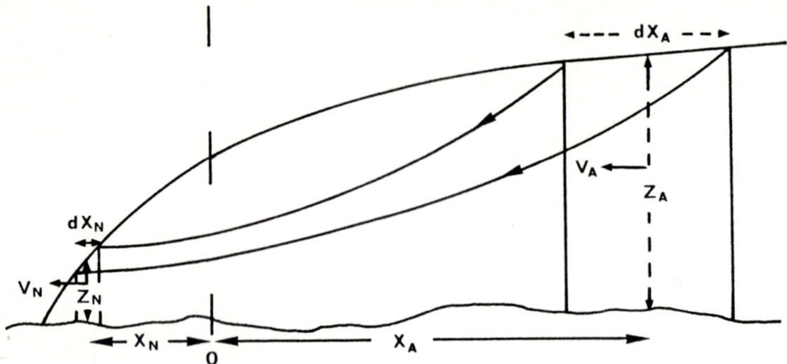

Fig. 3.11. Two-dimensional cross-section of an ice sheet showing symbols used for analysis of ice transport between pair of flow lines. (Robin 1983a, Fig. 1.20)

Ice velocities and ice thickness are also needed; their symbols are indicated in Fig. 3.11. For steady-state conditions to be maintained between the two flow lines, the mass transport of ice from upstream at x_A must equal the mass transport at x_N in the ablation zone. Therefore, Robin (1983a) gives the following equation for the steady state

$$\frac{dx_A}{dx_N} = \frac{N}{A} \cdot \frac{V_A}{V_N} \cdot \frac{Z_A}{Z_N} \; .$$

Table 3.3 gives the calculated and observed widths of ablation zone strips corresponding to accumulation zone sections, observed values being given by identical δ values. The general conclusion of the comparison is that excess ablation is observed in reality over that predicted by the steady-state model.

The sudden fall in δD values 200 m from the moraine to below the present-day δ values at Crête can be explained if the marginal ice had been deposited during the Last Glacial Maximum. A change of slope of the δ value versus distance relationship is displayed on Fig. 3.9 from Raynaud (1976). This probably reflects the change in δ values occurring at the boundary between Pleistocene and Holocene ice. Pleistocene ice is thus present at the surface of the ice sheet along the marginal zone in West Greenland.

Ice samples have been collected by Reeh and Thomsen (1986) from two profiles along the margin of Jakobshavn Isbrae, a major ice stream draining ice from the Greenland Ice Sheet a little south from the E.G.I.G. line (Fig. 3.12). Isotopic analysis of these samples gives generally higher δ^{18}O values in the down-glacier profile than in the up-glacier one, contrary to the trend expected in the ablation zone from Fig. 3.9. This probably indicates complex and non-stationary flow conditions in the marginal zone of the ice stream. It suggests, as quoted by Reeh and Thomsen (1986), that detailed isotopic studies can contribute to a better understanding of ice stream dynamics by distinguishing and locating the origin of the different types of ice composing the ice stream.

Table 3.3. Calculated and observed widths of ablation zone strips corresponding to accumulation zone sections, observed values being given by identical isotopic δD values. (Robin 1983b, Table 4.4)

Accumulation zone						Ablation zone					
Site	Distance from moraine (km) ($= X_A - 50$)	dX_A (km)	A (m/a)	V_A (m/a)	Z_A (m)	Corresponding points				dX_N	
						Distance from moraine (km) ($= X_N + 50$)	N (m/a)	V_N (m/a)	Z_N (m)	Calculated from equation (1.18) (km)	Observed (km)
Crête	510.8	122.7	0.32	6	3000	(0?)	2.7	18	100	1.31	0.2
Station Centrale	388.1	161.3	0.45	27	2700	0.2	2.7	20	250	1.84	3.3
Milcent	226.8	122.2	0.50	70	2050	3.5	2.5	27.5	500	2.34	9.0
Carrefour	104.6	54.6	0.40	104	1400	12.5	1.0	70	800	8.4	37.5
Equilibrium line	50.0					50.0					

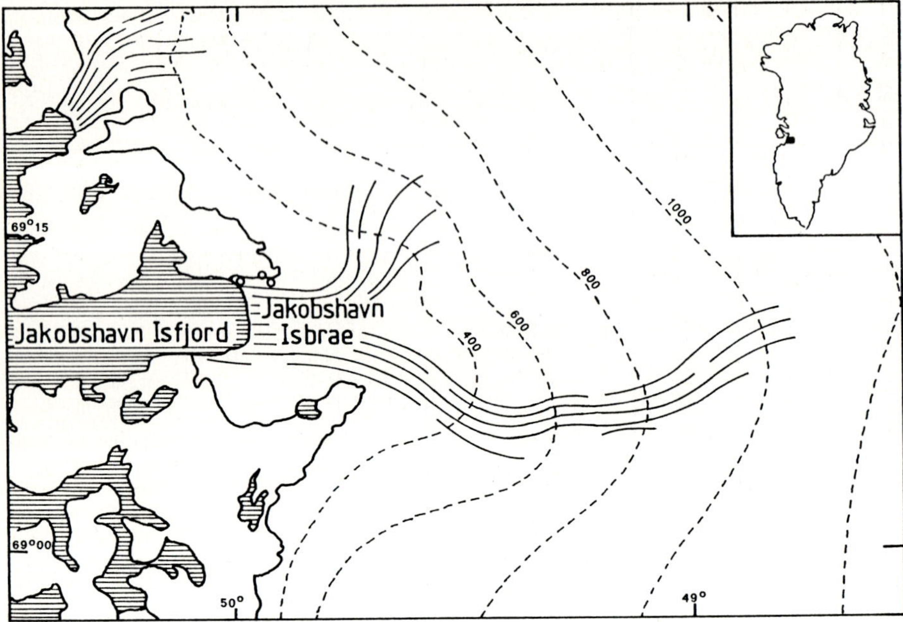

Fig. 3.12. Drainage basin of the Jakobshavn Isbrae. Locations for collection of ice and water samples are indicated by *open circles*. (After Fig. 1 in Reeh and Thomsen 1986)

Isotopic data from the coastal ice of Terre Adélie is displayed in Fig. 3.13. The deuterium content of two ice cores is given, along with that of the surface of the outer ablation zone: δD values are grouped in four classes such that the flow lines can be reconstructed. The most negative ice sampled originates about 1000 km from the coast. A complete δD profile is given for the drilled core G_1 in Fig. 3.14. It shows that at the extreme base of the profile, at an altitude below mean sea level, ice is found with very high δD values (around -50‰) which is probably of marine origin. This ice has a much lower gas content (1 to 2 cm^3/100 g of ice instead of 5 to 10 for the glacier ice above) and is a blue ice. The sodium content increases greatly in this blue ice to about 20 ppm, while the average for the profile is about 0.1 ppm. However, this amount of sodium is still low compared with that usually found in sea ice. Where freezing of sea water takes place in contact with the ice, very low concentrations have already been observed (Ragle et al. 1964). In the core G_2, such blue ice was not observed at the base, in this case located at an elevation of 7 m above sea level.

The relationship between structure and isotopic composition in the margin of the Barnes Ice Cap in Arctic Canada has been studied by Hooke (1976) and Hooke and Clausen (1982). Figure 3.15 gives the isotopic composition of the ice in $\delta^{18}O$ along the margin of the south dome of the Barnes Ice Cap. A prominent band of white, bubbly ice is displayed, which is both underlain and

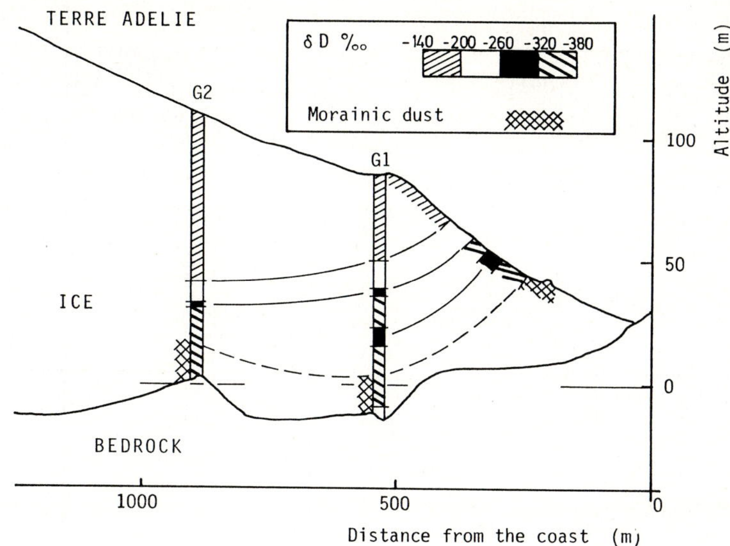

Fig. 3.13. The deuterium content in a cross-section of the coastal ice in Terre Adélie (Antarctica). (Lorius 1968, Fig. 1)

overlain by bluish-grey ice. $\delta^{18}O$ values are -38 to -40‰ in the white ice and -24 to -26‰ in the upper blue ice. This shift of about 14‰ is attributed to the climatic change between the Pleistocene and the Holocene. The lower bubble content of the overlying blue ice probably reflects an increase in the amount of percolating meltwater in the accumulation zone in the Holocene. This blue ice originally accumulated as superimposed ice under climatic conditions broadly comparable to those of the present day. Significantly, most of the accumulation on the Barnes Ice Cap today is in the form of superimposed ice. The blue ice beneath the white ice is also presumed to be of post-Pleistocene age. The normal stratigraphic order is reversed because this blue ice is superimposed ice which formed along the margin and which was subsequently overriden by the white ice during an advance of the glacier. Hooke (1976) gives a sketch of this evolution reproduced in Fig. 3.16. Sometimes the white ice layer is present in two zones along the margin, the repetition believed to be the result of folding. Figure 3.17 from Hooke and Clausen (1982) shows the situation with the corresponding $\delta^{18}O$ values. Hudleston (1976) has demonstrated that such folds may form when minor advances or retreats of a glacier change the pattern of flow lines over irregularities in the bed.

Reeh and Thomsen (1986) have also studied the $\delta^{18}O$ in surface ice from the part of the Greenland Ice Sheet draining to Pakitsup, north of Jakobshavn Isbrae along the GGU ablation-stake line. Figure 3.18 from Reeh and Thomsen (1986) shows the regression obtained between the $\delta^{18}O$ content of the meltwater samples and elevation. Another straight line expresses the linear regression between elevation and snow and ice δ values from the accumulation

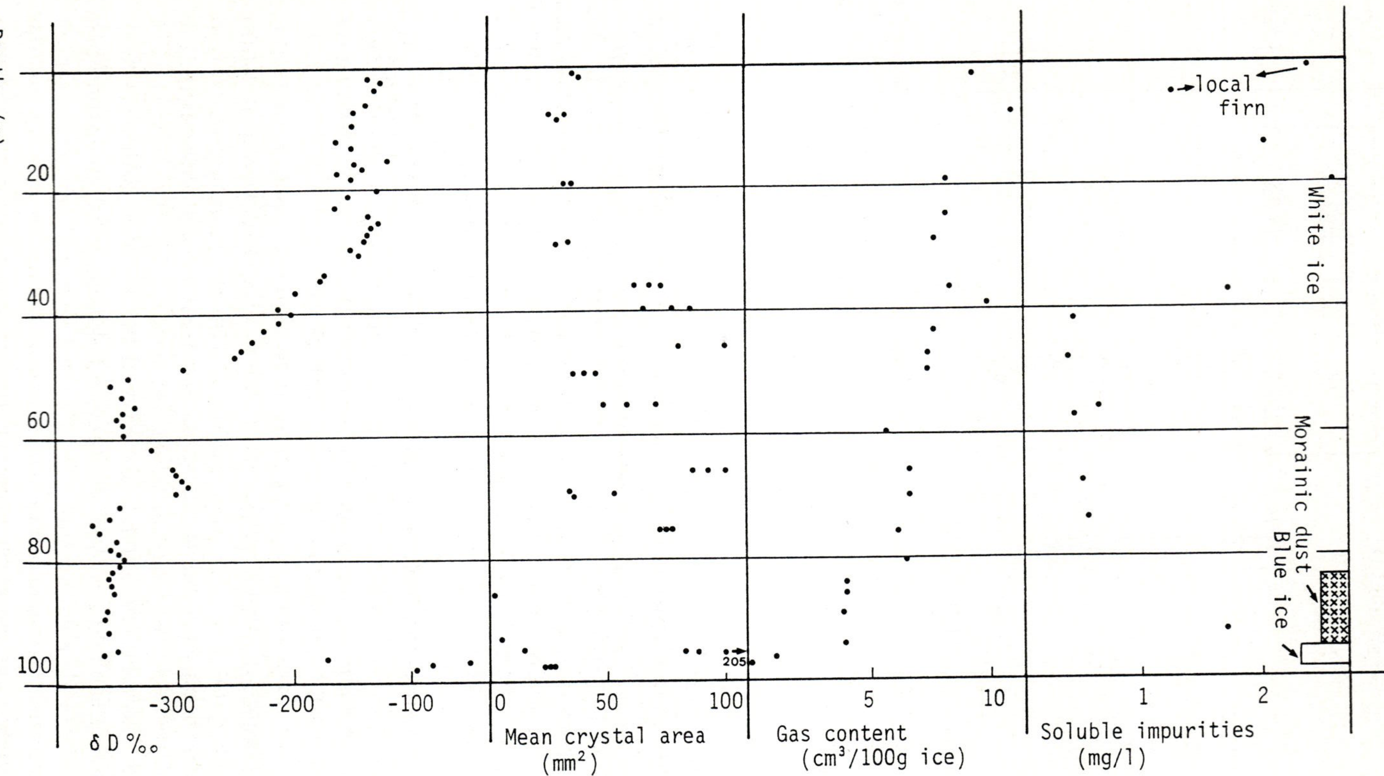

Fig. 3.14. Physical and chemical properties of ice from the drilled core G1 in Terre Adélie (Antarctica). (Lorius 1968, Fig. 2)

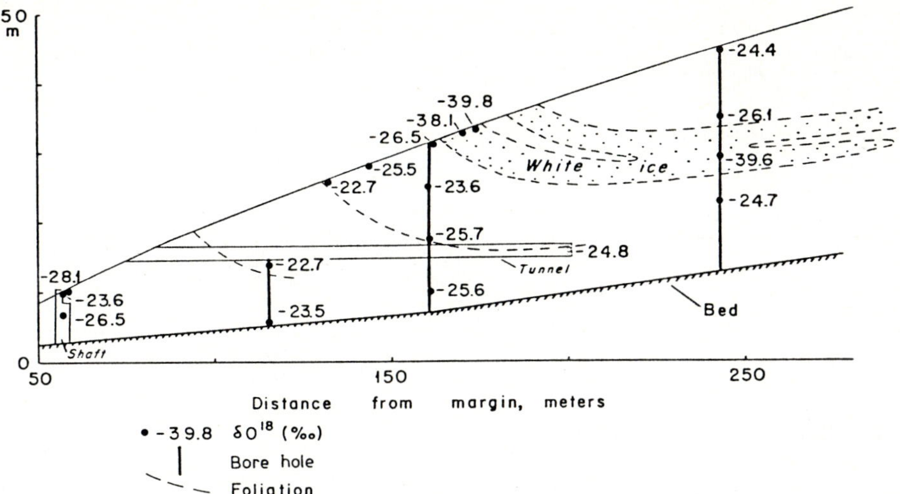

Fig. 3.15. Relationship between structure and isotopic composition in the margin of the Barnes Ice Cap. (Hooke 1976, Fig. 2)

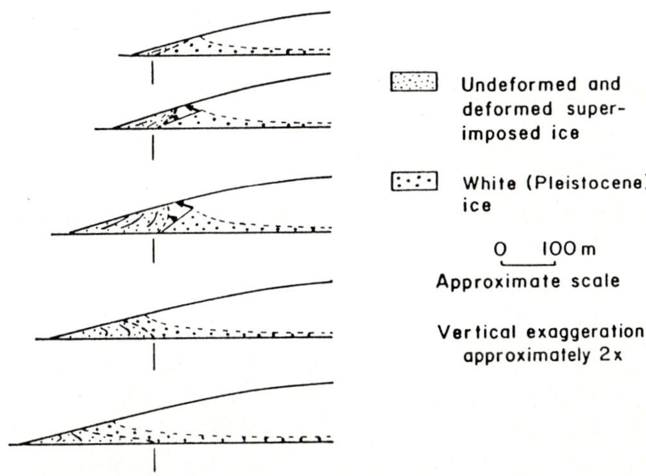

Fig. 3.16. Sequential cross-sections showing schematically, *from top down*, the process by which superimposed ice is believed to be overridden during an advance of the glacier. Such overriding occurs if the glacier is frozen to its bed, because velocity increases with height above the bed. *Light solid lines* in superimposed ice depict schematically the progressive overturning of original sedimentary banding. In nature such banding is commonly obscured by development of foliation in later stages of deformation. (Hooke 1976, Fig. 3)

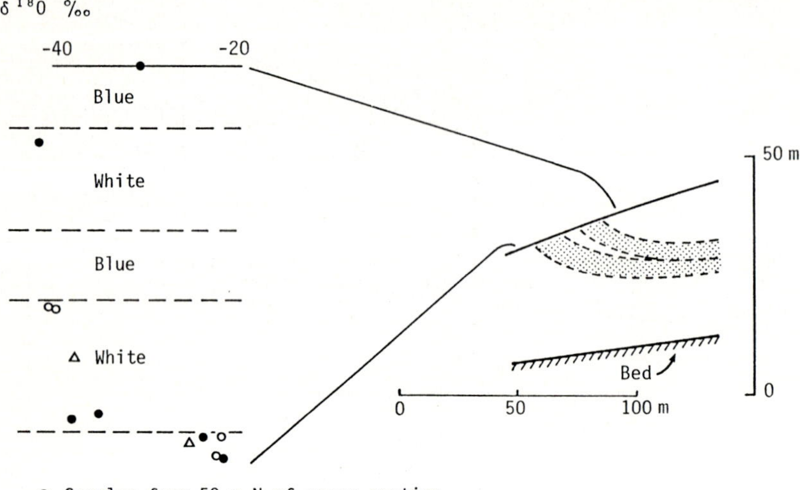

Fig. 3.17. $\delta^{18}O$ variations in ice near "Camp Bruce", Barnes Ice Cap. Repetition of white ice layer is due to folding. (Hooke and Clausen 1982, Fig. 3)

area together with winter snow δ values from the ablation zone. The first line is given by:

$$\delta^{18}O = -33.6 + 0.0113\ E\ ,$$

where $\delta^{18}O$ is expressed in permil and E is the elevation in metres and the second by:

$$\delta^{18}O = -15.1 - 0.005\ E\ .$$

The intersection of the two lines defines the elevation of the equilibrium line. This gives an elevation of about 1100 m, in agreement with direct observations in the field. The outermost part of the ablation zone is probably Pleistocene ice with values around $-30‰$ in $\delta^{18}O$. However, ice from the 30 m directly adjacent to the margin shows $\delta^{18}O$ values in striking contrast with the Pleistocene ice with mean values clustered around $-20‰$ in $\delta^{18}O$. Samples of local isolated firn patches give $\delta^{18}O$ values similar to those of this very marginal ice. Reeh and Thomsen (1986) considered these rather high values as evidence that this ice originates from local wind-drift snow. This snow accumulates along the ice margin during the winter and is transformed into superimposed ice during the summer melt season. The sequence displayed near Pakitsup, north of Jakobshavn Isbrae, is thus to some extent similar to the one described by Hooke for the margin of the Barnes Ice Cap. The isotopic study of surface ice samples from an ice sheet can thus provide useful information relating to ice flow.

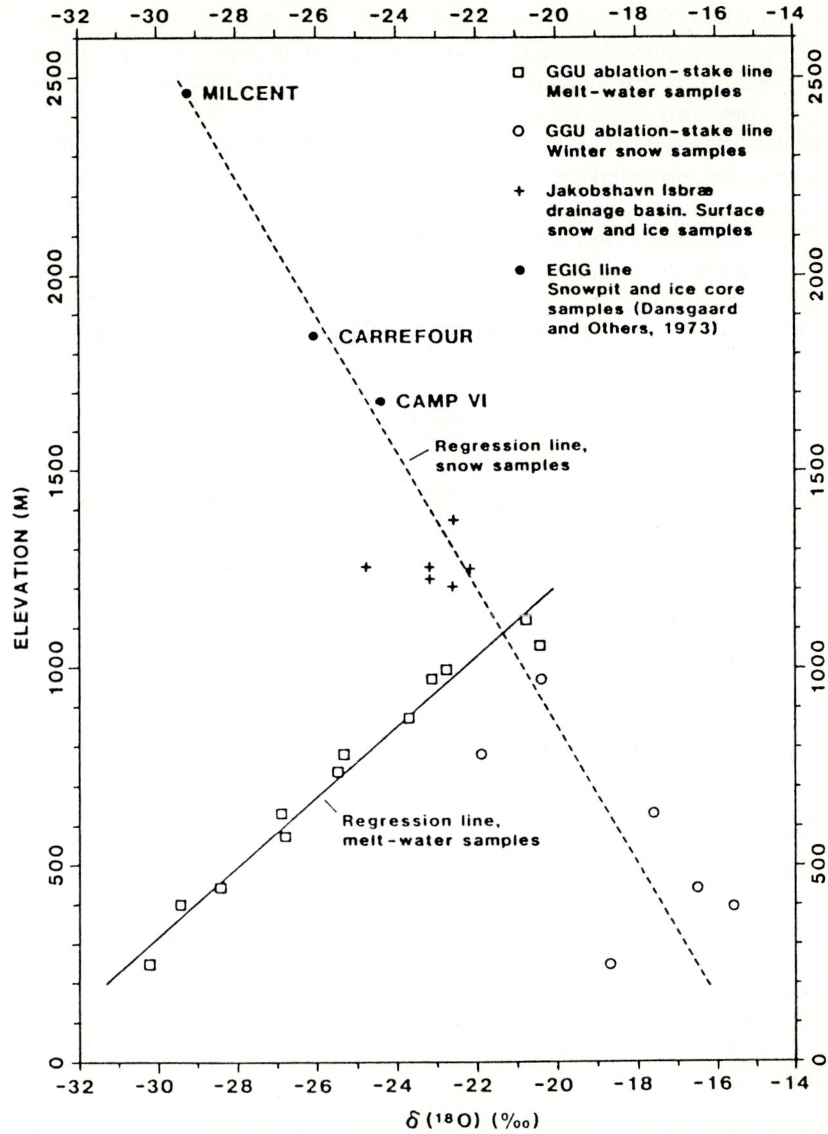

Fig. 3.18. $\delta^{18}O$ in surface ice and snow samples from the Greenland ice-sheet sector draining to Pâkitsup akuliarusersua (N.E. of Jakobshavn), plotted versus elevation. (Reeh and Thomsen 1986, Fig. 3)

3.4 Lead 210 in Ice and Alpine Glacier Flow

The ^{210}Pb ice dating method enables the reconstruction of flow lines for a specific glacier if sufficient dates are available. However, the absence of melting and of meltwater percolation seems to be a requirement for the use of the

^{210}Pb dating method. This is due to the fact that, in the presence of water, snow and firn do not behave as a closed system for ^{210}Pb subsequent to snow deposition. In alpine glaciers where ice exists at the melting point, there is little hope that this requirement is met for the firn. However, ice itself can behave as a closed system, as indicated by a combined study of ^{90}Sr and ^{210}Pb measurements, reported by Picciotto et al. (1967) for Kesselwandferner, a glacier in the Austrian Alps. The reason for this combined approach was the hope that ^{90}Sr, a radionuclide produced by nuclear bombs, would provide some information on the extent of chemical exchanges in the firn and ice layers.

Figure 3.19 gives ^{90}Sr activity in dpm/kg (disintegrations per minute per kilogram of ice) with depth in the firn layer of Kesselwandferner. This profile

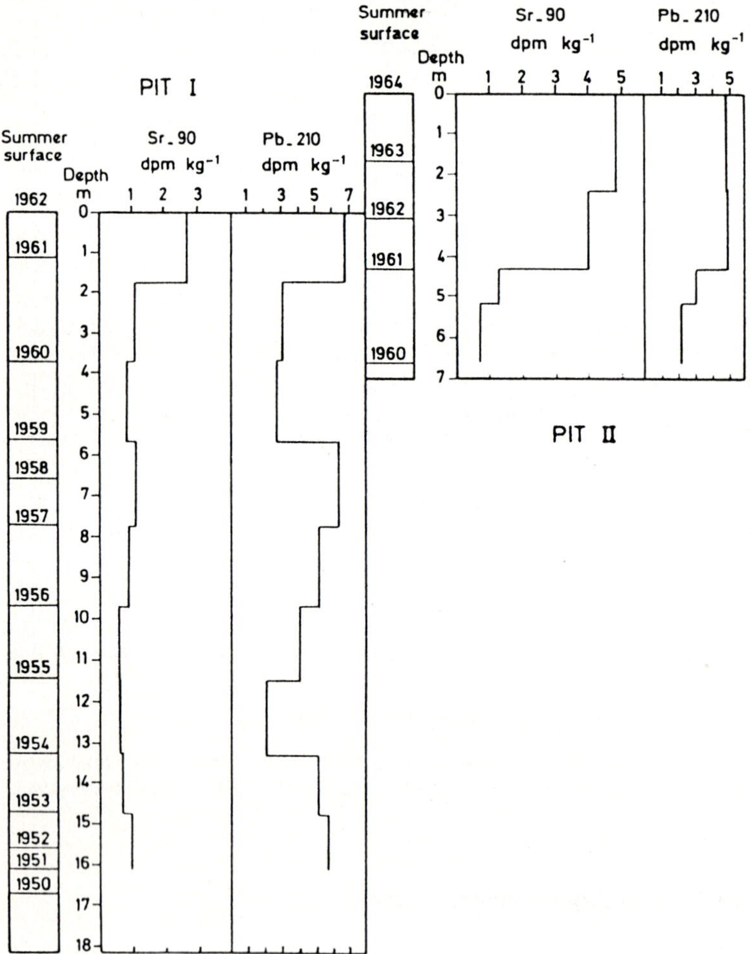

Fig. 3.19. ^{90}Sr and ^{210}Pb specific activity versus depth in the firn of Kesselwandferner, Austrian Alps. (Picciotto et al. 1967, Fig. 2)

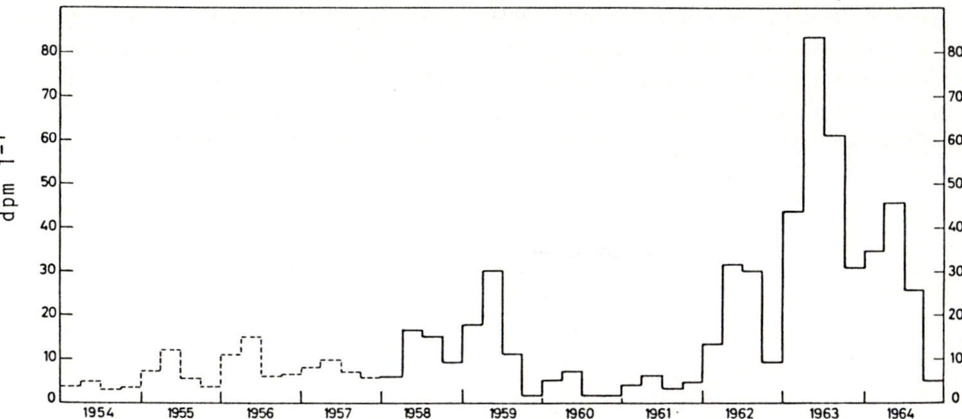

Fig. 3.20. ^{90}Sr activity in the precipitations. *Solid line* average of Vienna and Klagenfurt. *Broken line* average for the latitude band 30–60°N. (Picciotto et al. 1967, Fig. 3)

is compared with Fig. 3.20 representing the expected variation of the ^{90}Sr content in the precipitation. The ^{90}Sr activity of the firn shows little variation with depth, with the exception of a well-marked increase above the 1961 summer surface, reflecting the resumption of the nuclear tests in October 1961. However, the ^{90}Sr in the firn is an order of magnitude lower than its average content in the precipitation, illustrating the intensity of chemical exchanges within the firn. Now, the absence of ^{90}Sr in ice formed from snow precipitated well before 1952 indicates a lack of exchange with recent meltwater or rainwater. Figure 3.19 also gives the ^{210}Pb specific activity variations down the firn

Table 3.4. ^{210}Pb in near-surface ice samples from the ablation area of Kesselwandferner, Austrian Alps. (Picciotto et al. 1967, Table 2)

Sample number[a]	^{210}Pb concentration (dpm/kg)	^{210}Pb age (years)
1	2.4 ± 0.1	18
2	3.2 ± 0.2	9
3	1.3 ± 0.1	38
4	1.61 ± 0.05	31
5	0.52 ± 0.03	67
6	0.53 ± 0.04	66
7	0.21 ± 0.03	95
8	0.70 ± 0.05	57
9	0.27 ± 0.03	88
10	<0.05	>150
11	<0.03	>150
12	<0.05	>150
13	<0.04	>150

[a] See position on map, Fig. 3.21.

profile. In contrast with the ^{90}Sr results, the values of the specific activity for ^{210}Pb are close to those expected in the precipitation. There is thus a striking difference between the behaviour of ^{90}Sr and of ^{210}Pb in the firn which could be due to differences in their chemical properties. The ^{210}Pb activity of the ice at the time of its formation seems to be equal to the average activity found in

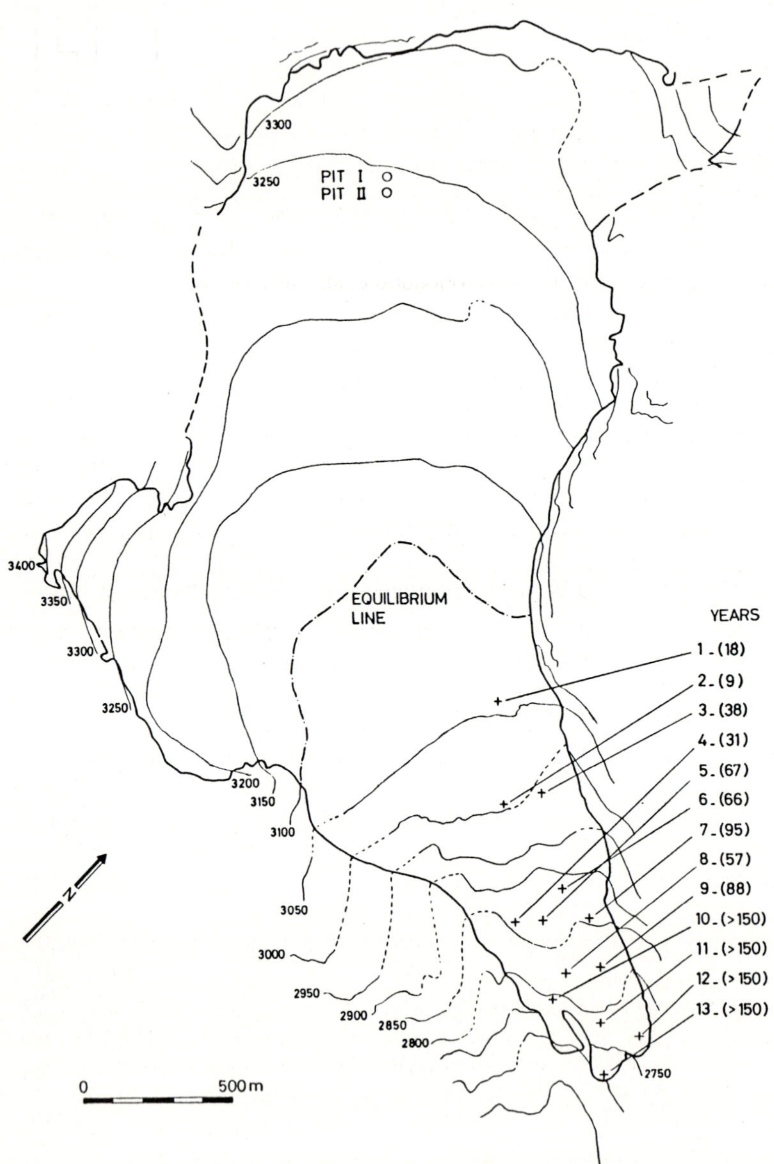

Fig. 3.21. Sketch map of the Kesselwandferner, Austrian Alps. Location of the samples. *1* Sample number: *(18)* ^{210}Pb "age" in years. (Picciotto et al. 1967, Fig. 1)

the firn, namely 4.3 dpm/kg. The possibility of the artificial production of ^{210}Pb by nuclear test bombs in the atmosphere cannot be ruled out, but it is unlikely that this significantly affects the values obtained in the precipitation. Furthermore, the distribution of ^{210}Pb activity in the surface ice is in qualitative agreement with the distribution of ice, by age, expected from considerations of classic glacier dynamics. Thus, once formed, ice behaves as a closed system for ^{210}Pb, implying that the variations in ^{210}Pb content are due solely to radioactive decay and not to chemical exchanges. The ages derived from the law of radioactive decay would represent the time elapsed since the transformation of firn into ice, with zero age taken at an activity of 4.3 dpm/kg of ice, the mean value of the firn. These are given in Table 3.4 and in Fig. 3.21. The ^{210}Pb dating method applied to alpine glaciers will thus not give, as in polar areas, the time elapsed since the deposition of the snow.

Figure 3.21 shows the age distribution of surface ice samples from Kesselwandferner. The general increase in age as one proceeds from the equilibrium line of the glacier to the snout is well displayed. This is connected with the upward component of velocity in the ablation zone. At the glacier snout, the age is greater than 150 years, i.e. beyond the range where the method is capable of precise dating. By considering the group of points formed by samples 4, 5, 6 and 7 it can be assumed that the 65-year surface isochron follows approximately the 2875-m contour line. When compared with a measured present-day surface velocity of 20 m/year, the agreement is satisfactory considering the distance from the sampling site to the equilibrium line and possible flow lines configurations. The internal consistency of the ages obtained supports the validity of the ^{210}Pb dating method in alpine glacier studies. Although the ice samples from the snout are too old to be dated by this method, they are probably close enough to 150 years to attest to the short ice residence time in this Alpine glacier. With the oldest ice located at the snout of an alpine glacier, dates from the frontal zone can give an idea of the ice residence time, as indicated in Section 1.4. This makes it possible to determine the response time of a specific alpine glacier to a change in mass balance due to climatic effects.

3.5 Mineral Particles in Ice and Glacier Flow

The debris transported by glaciers is derived from two principal sources: supraglacial and subglacial. Their subsequent transport paths are illustrated in Fig. 3.22. If the supraglacial debris is deposited on the glacier surface in the ablation zone, it will remain on the surface. If the supraglacial debris is deposited on the glacier surface above the equilibrium line, the transported debris is buried by snow and, as the vertical component of the velocity vector is directed downwards in the accumulation zone, will follow an englacial path. This debris can regain a supraglacial position downglacier with the emergence of the flow line in the ablation zone. In the case of a valley glacier, debris falling onto the glacier surface from the headwall will tend to travel immediately

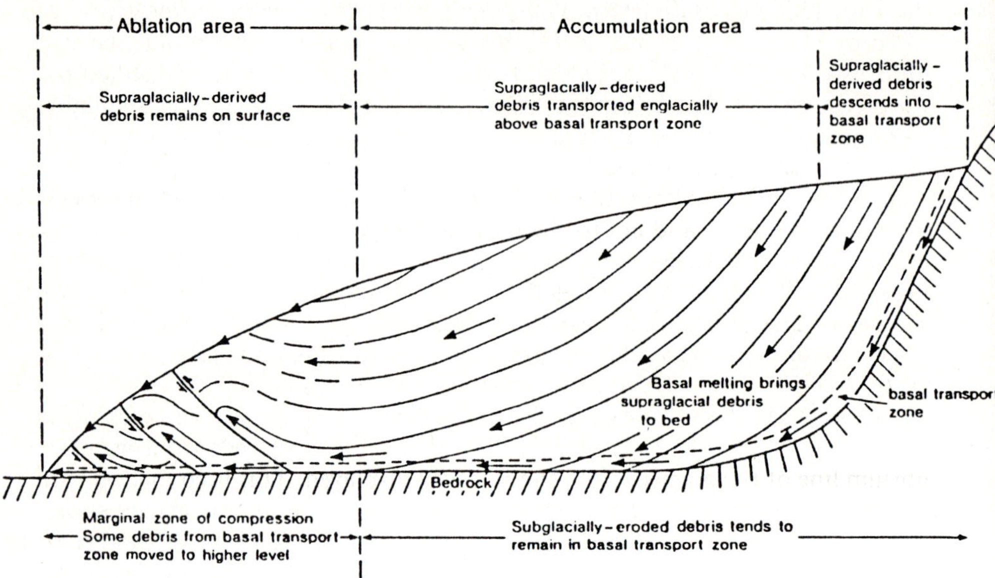

Fig. 3.22. Transport paths of debris in a cirque glacier or valley glacier derived from a headwall. This model can also be applied to an ice cap pierced by nunataks or outlet valley glaciers draining an ice cap. (Boulton 1978, Fig. 1)

above the glacier bed as the glacier margin forms one extremity of a basal flow line (Boulton 1978). Because it is incorporated in the glacier adjacent to the headwall, this debris will, initially at least, travel in the basal transport zone of Fig. 3.22. Debris falling onto the lateral margins of a valley glacier above the equilibrium line will also tend to be transported immediately above the glacier bed and will additionally tend to move towards the centre line of the glacier due to the pattern of flow lines in the accumulation zone. However, debris falling onto the lateral margin of a valley glacier below the equilibrium line is unlikely to be transported far from the source area since flow lines are directed outwards in the ablation zone. A special case, however, is met when rock avalanches or debris flows reach the central part of the glacier by displacement on its surface away from the margin.

Basal melting may cause englacial debris to reach the glacier bed. In glaciers at the pressure melting point in their lowermost part, the basal transport zone is restricted to a thin layer (up to 1 m) of debris-rich ice immediately above the glacier sole. The vertical dispersion of subglacial debris tends to be restricted by the mechanism of pressure-melting and regelation which is of importance only for small bed protuberances. In polar glaciers, subglacially derived debris may be lifted to a relatively high level of englacial transport, several tens of metres above the glacier bed, possibly as a result of net freezing of meltwater to the glacier sole along with subsequent folding of the ice or displacement along shear zones or shear planes. In some alpine glaciers, this

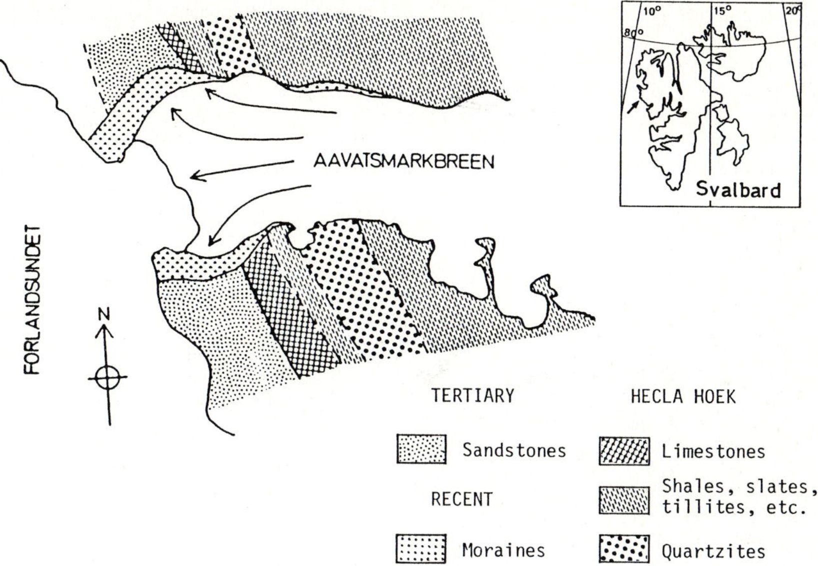

Fig. 3.23. Generalized and very approximate map of lithologies over which Aavatsmarkbreen flows. (After Fig. 6 in Boulton 1970)

phenomenon of net basal ice accretion is also present, but does not appear to be of the magnitude observed in numerous polar areas.

The hypothesis of basal freezing implies that, with the exception of localized areas, conformable debris sequences are produced, a contention tested by Boulton (1970) at Aavatsmarkbreen, Oscar II Land, Svalbard. This glacier flows across different bedrock lithologies disposed at approximately right angles to the general horizontal ice velocity component. Shales outcrop in the upper glacier basin; they are followed downglacier by a band of quartzites, then by a thin band of shales. In the lower glacier basin, a limestone band is followed downglacier by a band of sandstones. The different lithologies and the glacier position are sketched in Fig. 3.23. The frontal ice cliff of Aavatsmarkbreen has been examined. Table 3.5 gives the vertical variation in englacial debris types. The upward sequence reflects, in reverse order, the lithologies over which the glacier has moved: recent unlithified marine sediments and Tertiary sandstones and Hecla Hoek rocks, in which some of the prominent lithologies are recognizable. This sequence thus agrees with that predicted by a basal freezing hypothesis. The anomalous position of the specimens of Tertiary sandstone at 8.5 m could be explained by movement along some cross-cutting structure such as a thrust plane.

Boulton (1970) has made the generalization that polar glaciers carry considerable amounts of englacial debris derived from the glacier bed, whereas temperate glaciers carry very little basally derived englacial debris. This apparent contrast supports the suggestion that temperature regime is the controlling

Table 3.5. Vertical variation in englacial debris types on Aavatsmarkbreen, Spitsbergen. (Boulton 1970, Table 1)

Height of sample above basal thrust plane (m)	Number of samples of each debris type						
	Tertiary sandstones	Limestones	Massive quartzites	Shales, slates, etc.	Tillite	Granite, vein quartz, etc.	Totals
9.5				12	2	6	20
9.0				9	2	3	14
8.5	4		2	7	3	4	20
8.0			10	4	2	2	18
7.5			8	7	3	2	20
7.0		9	5	3			17
6.5	2	14	3	4		3	26
6.0	21	2	3	5		2	33
1.5 – 5.5	Recent marine sediments						

factor in the englacial incorporation of large volumes of basally derived debris and that only those glaciers in which basal freezing is an important process show large amounts of such debris. This carries the corollary that glaciers which are entirely frozen to their beds will contain very little englacial debris. Basal freezing can be shown to have occurred by studying the isotopic composition of basal ice, as developed in the following chapters of this book. The generalization made by Boulton has, however, been rejected by Andrews (1971, 1972). Clapperton (1975), who has examined subpolar glaciers in Svalbard and a temperate glacier in Iceland in the context of their periodic surging, has argued that such behaviour is of much greater significance to the Boulton-Andrews controversy than the generalized thermal characteristics of the glaciers. There is thus a possibility that a causal relationship exists between surging behaviour and high subglacially derived debris contents. Moreover, basal meltwater plays a prominent role in surging while glaciers which are dry-based contain very little subglacially derived debris in the absence of basal meltwater. In surging glaciers, cavities filled with low pressure water may exist where freezing and debris incorporation can occur. But observations indicate that high water pressures in surges and sliding friction lead to very large basal melting. Also bands are too extensive to come from single cavities (Sharp 1985). The fact that the surge sometimes propagates the glacier over unconsolidated sediments existing in the frontal zone is also of importance. Such sediments are readily susceptible to being picked up and this contributes to the high englacial debris contents of surging glaciers. However, the role of compressive tectonic processes in front of propagating surge fronts in generating rich zones of debris-rich ice at glacier terminus must also be considered.

The small Sydkap Ice Cap, situated in South-Western Ellesmere Island (Arctic Canada) has been investigated by Souchez (1971) in the context of the use of debris lithology to reconstruct characteristics of glacier flow. In particular, an ice lobe on its north-west margin has been studied and is shown in

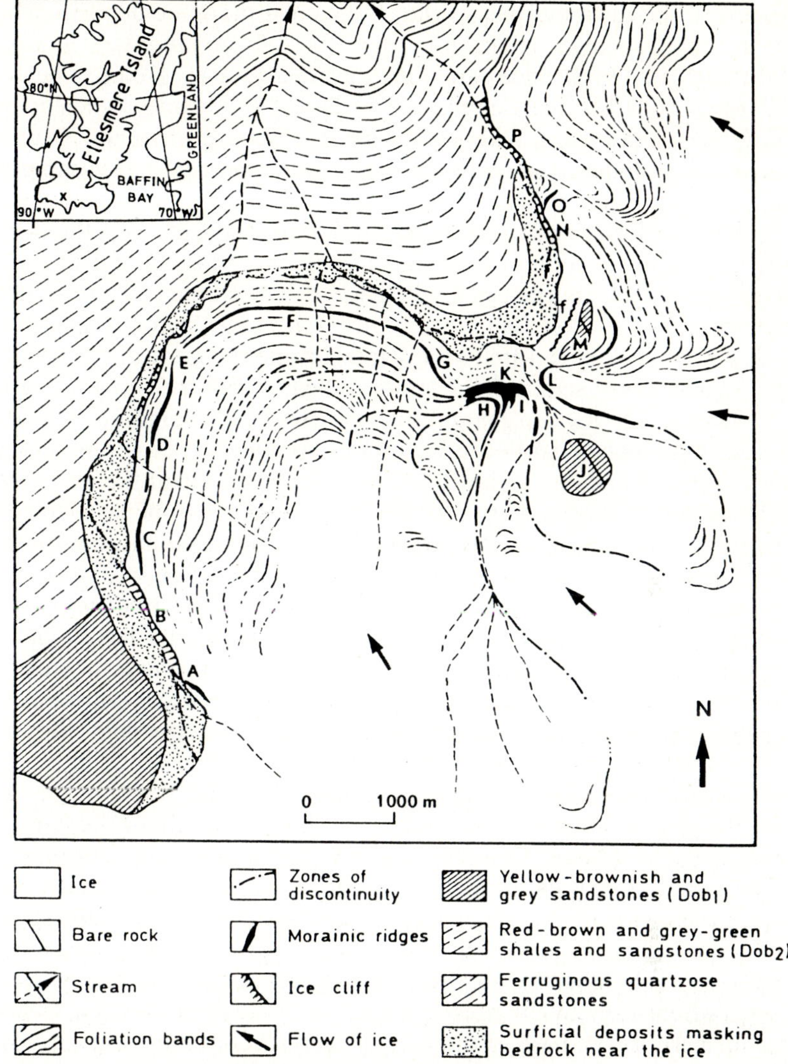

Fig. 3.24. The marginal zone of the ice lobe investigated in the north-western part of Sydkap Ice Cap, Ellemere Island, Canada. (After Fig. 1 in Souchez 1971)

Fig. 3.24. This lobe is fringed by an ice-perched moraine, also called an ice-cored moraine, following Østrem (1965), who coined the term. It covers different outcrops of Devonian sedimentary rocks gently dipping to the north-north-west. Of particular interest is the limit between red-brown and grey-green siltstones (Dob 2) and yellow-brownish and grey sandstones (Dob 1). This limit can be traced with confidence beneath the ice. The source area of material in the ice and in the moraines can thus be located. At locality D, outer and inner zones can be distinguished in the ice-perched morainic deposits which necessarily have a subglacial origin, since there is no nunatak on this ice cap. Lithologically, these zones are quite different (gravel counts). The debris from the outer zone is largely composed of Dob 1 sandstones, whereas that of the inner zone has a much higher proportion of Dob 2 siltstones. Thus rock fragments of the inner zone derive from a source nearer to the ice cap margin than those of the outer zone.

Another feature of the ice-cored moraine at locality D is the occurrence in the inner zone of an unusually large boulder (15 m long, 5.5 m wide and 7 m high), elevated from the bed of the ice cap. The contact between the two lower members of the Okse Bay Formation (Dob 1 and Dob 2) is preserved in this block which has been lifted as a whole despite the great differences in lithology. The observed sequence of strata indicates that the block is right side up.

Between the ice cap margin and the ice-cored moraines at locality D, debris-rich ice bands can be seen dipping up-glacier. The lithological composition in these bands is similar to that of the outer zone of the ice-cored moraine (Table 3.6). It thus appears that the inner ice-cored zone at locality D derives its debris from a point nearer to the ice cap margin than does the outer zone or the ice between the outer zone and the margin. Basal ice layers bearing Dob 1 sandstone fragments and having an upward component of velocity in the ablation zone are intersected by an active shear zone transferring debris from a source nearer to the margin to the ice surface and forming the inner zone rich in Dob 2 siltstone fragments. The large piece of bedrock must have been displaced along the same active shear zone. This example shows that the lithological composition of englacial and supraglacial debris in ice can be used as a tracer of specific dynamic processes of glacier flow.

Hooke (1973 b) has studied the flow near the margin of the Barnes Ice Cap, Baffin Island (Arctic Canada), and the development of ice-cored moraines. Debris-rich ice is present at the base along the margin of the ice cap but, beneath the debris bands, clean blue ice appears. This is deformed superimposed ice produced at the margin by the refreezing of percolating meltwater in drifting snow, subsequently overriden during an advance of the glacier. Because longitudinal flow in the ablation zone is usually compressive, debris-bearing ice formerly at the base of the ice cap moves upward and outward over the superimposed ice during such an advance. The situation is shown in Fig. 3.16: the debris-bearing ice is at the base of the white, bubbly Pleistocene ice while the clean blue ice beneath is the deformed superimposed ice. Temperature distribution in the margin of the Barnes Ice Cap (Fig. 3.25) gives additional information in this context. Latent heat released by the refreezing of meltwater to form the superimposed ice has an appreciable effect because it reduces the amplitude of the winter cold wave. The temperature distribution suggests that superimposed ice increases in importance abruptly at about 150 m from the margin. The lithological contrast of the two ice-perched morainic zones in the north-western part of Sydkap Ice Cap implies peculiar flow behaviour along the margin and the existence of deformed superimposed ice cannot be ruled out. Further study must be carried out to assess whether an active shear zone

Table 3.6. Lithological composition of debris in and on the ice of Sydkap Ice Cap, Ellesmere Island, Arctic Canada. (Souchez 1971, Table 2)

		Base of ice cliff north of locality B[a]		Foliation bands between the ice-cap margin and the moraines at locality D[a]		Ice-perched morainic outer zone at locality D[a]			
Diameter		12 – 25 mm	6 – 12 mm	12 – 25 mm	6 – 12 mm	12 – 25 mm		6 – 12 mm	
Red-brown siltstone	Dob_2	4%	8.5%	5%	9%	4%	5.6%	6.3%	9.5%
Grey siltstone	Dob_2	22%	22.9%	12%	18%	–	7.8%	10.0%	16.4%
Yellow sandstone	Dob_1	42%	37.3%	54%	51%	60%	67.7%	43.0%	52.2%
Grey sandstone	Dob_1	32%	31.3%	29%	22%	36%	18.8%	40.7%	21.9%

[a] See Fig. 3.24 for location.

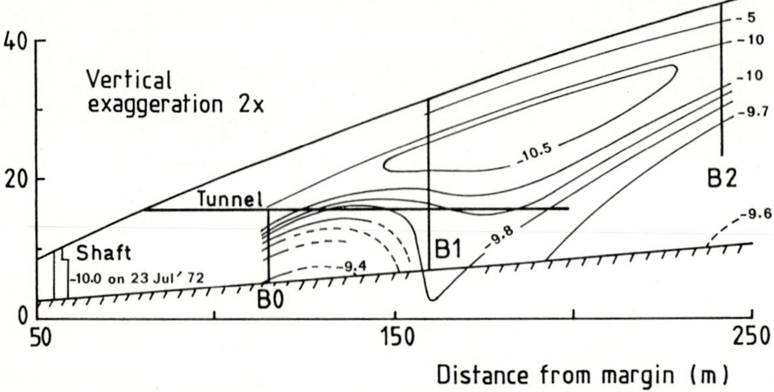

Fig. 3.25. Temperature distribution along a pole line in the margin of Barnes Ice Cap; August 20–23, 1970. *Dashed contours* extrapolated. (Hooke 1973b, Fig. 5)

is responsible for the formation of the inner ice-cored moraine or, as Hooke suggested, whether it could be a product of changing thermal regimes at the base of the glacier entailing a down-glacier shift in the location of maximum subglacial erosion.

3.6 Evidence for Buried Glacier Ice

Outside the present-day limit of ice caps and ice sheets, the ground may be perennially frozen and may therefore contain buried ice masses. Indeed, buried glacier ice from a former extension of an ice cap or an ice sheet is likely to occur in such circumstances, and its existence testifies to former ice flow in a glacier with a different surface profile from that of the present.

Most authors who have investigated large ground ice masses consider them to be segregation features (Mackay 1971; Mackay and Black 1973; French 1976). The term segregation ice corresponds to ground ice that is formed at the freezing front when freezing propagates into water-rich sub-permafrost material. This water can be sucked towards the freezing front to give rise to lenticular masses of pure ice. However, in certain cases, burial of surface ice – be it glacier ice or river ice – cannot be excluded. The following example demonstrates that the study of the composition of such ice may allow discrimination between possible origins.

In the Prince Albert Peninsula on northern Victoria Island in Western Arctic Canada, many sites are underlain by large bodies of ground ice sometimes exposed at the base of headwalls by active slumps. Numerous slumps visible in the area are, however, presently inactive because the retreat of their scarp faces has stopped, and in such circumstances, ice is no longer visible at the base. Two slumps have been studied on the west side of Richard Collinson Inlet by Lorrain and Demeur (1985); they are indicated in Fig. 3.26. They have a

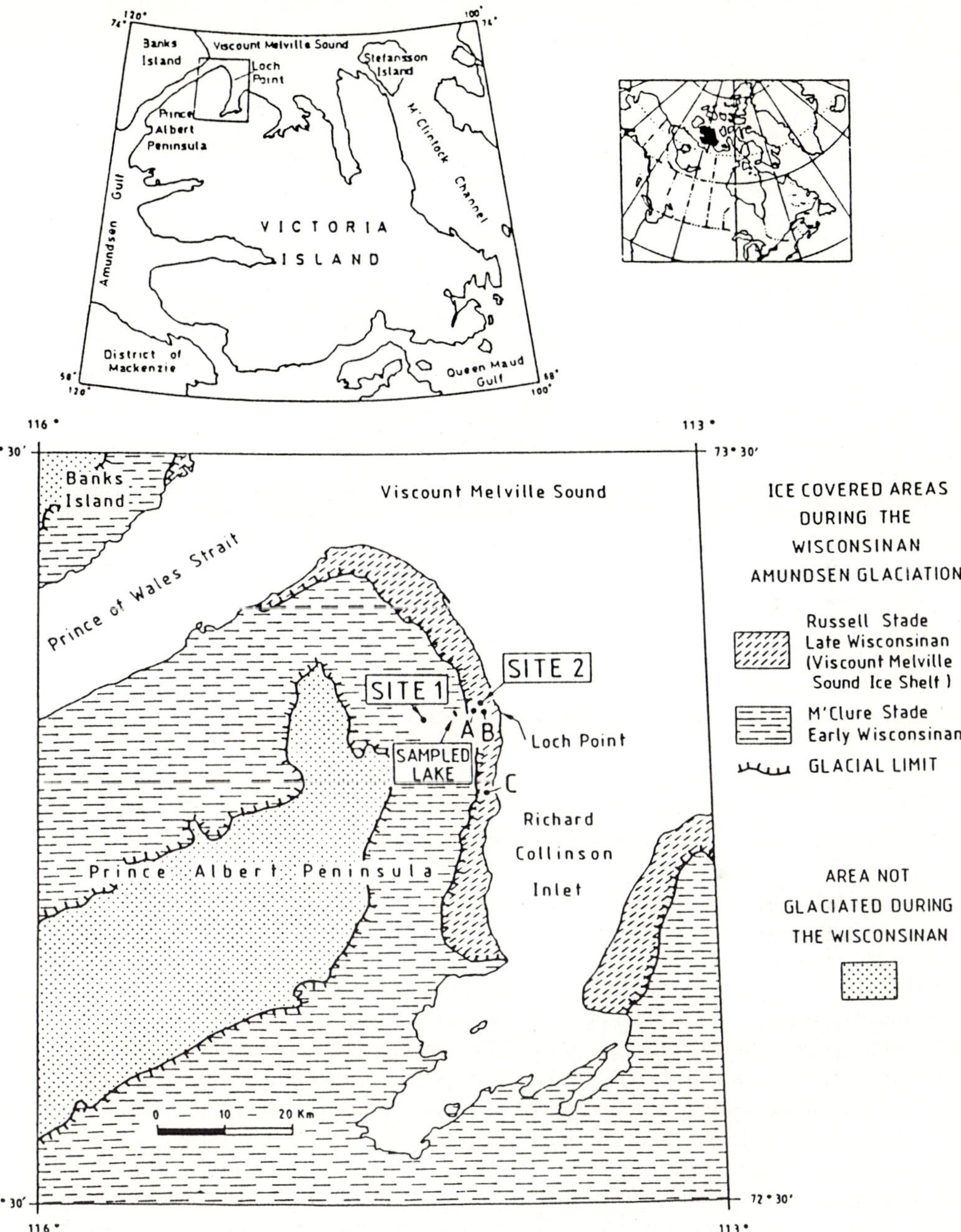

Fig. 3.26. Location maps of the investigated sites on Victoria Island (glacial limits after Hodgson and Vincent 1984). (Lorrain and Demeur 1985, Fig. 1)

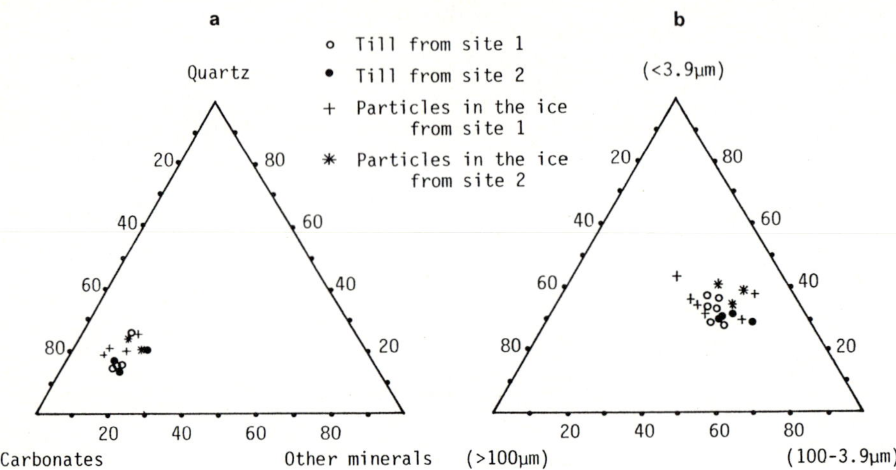

Fig. 3.27. Characteristics of sediments from the ice and from the overlying tills at the sites investigated on Victoria Island. **a** Mineralogical composition. **b** Grain-size distribution. (Lorrain and Demeur 1985, Fig. 5)

slump face that can be divided into three sections: a vertical one at the top, corresponding to a till about 1 m thick; a fairly regular slope with gradient varying between 20 to 35°, corresponding to the ice; and a scarp base, with a very weak slope corresponding to the accumulated slumped debris. The ice shows massive, subvertical banding. The banded appearance is due to an alternation of not only debris-rich and relatively clear ice, but also of bubbly and bubble-free ice. At one site, an ice wedge is present within the ice body. This ice wedge showed an alternation of layers less than 1 cm thick with different amounts of mineral and air inclusions with an irregular banding. The most striking feature of these layers was that they frequently criss-crossed each other. This facies was visible over a maximum width of 3 m, distinctly different from the enclosing ice mass.

Characteristics of sediments from the ice and from the overlying till are compared in Fig. 3.27. The particle content of the ice masses is lower than the 30% value currently cited as normal for segregation ice (French 1976). Particles in the ice and in the till appear to be indistinguishable from one another: both have the same grain size distribution and the same mineralogical characteristics. This does not discount a glacial origin for the buried ice, since the overlying till has most probably been carried into the area by the former ice movement of the same glacier.

Isotopic composition of the ice, both in δD and in δ^{18}O, illuminates the problem further. Figure 3.28 represents the δD-δ^{18}O diagram on which all the samples analyzed were plotted. The ice samples form two groups on the diagram. The less negative group corresponds to samples from the above- mentioned ice wedge, while the second group is composed of ice samples from the large ice bodies. The latter are well aligned on the δD-δ^{18}O diagram with a

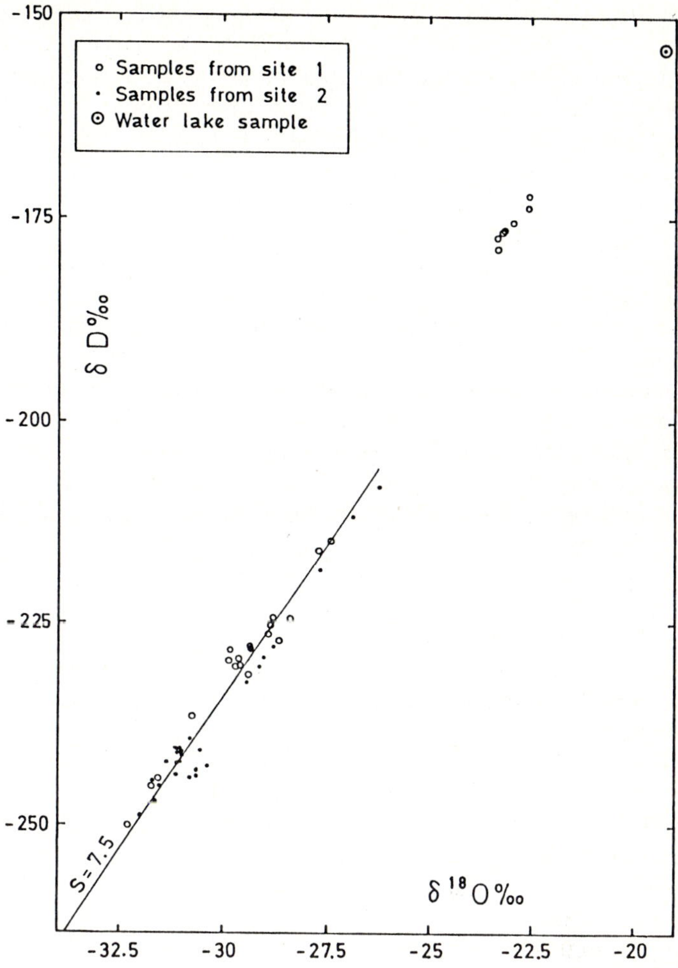

Fig. 3.28. The isotopic composition in δD and δ^{18}O of the ground ice samples and of a nearby lake water sample from Victoria Island. *S* is the slope of the linear regression line of the ice samples except those taken from the ice-wedge. (Lorrain and Demeur 1985, Fig. 6)

slope S of 7.5 and a correlation coefficient is 0.96. This linear relationship is attributable to the precipitation effect defined by Craig (1961). A slope of 8 is considered as a worldwide mean value, but slopes of 7.5 have been reported in precipitation at some circumpolar stations by Dansgaard (1964), among others. This relationship is also considered as valid for glacier ice. In contrast, ice formed by freezing of a water body will display points aligned along a distinctly lower slope called a freezing slope, as seen in Chapter 2. Also, in certain circumstances, constant δ values can be obtained. Such a constancy has been observed in the case of an ice core retrieved from the frozen sediments in the bottom of an artificially drained lake in which the freezing front was still migrating downwards at the time of drilling (Michel and Fritz 1982). The mean value in δ^{18}O of the ice samples from the bodies studied by Lorrain and Demeur (1985) were about 11‰ more negative than surface water samples which correspond to values of present-day precipitation. This shift is the same as that found by Dansgaard et al. (1971) in the Camp Century ice core and is interpreted as marking the transition between the Last Glacial Maximum and the Holocene. However, this climatic amelioration is responsible for a δ^{18}O shift of approximately 7‰, the remaining part being related to the elevation effect. These results indicate that buried Pleistocene glacier ice is present on Northern Victoria Island and that ice wedges such as the one mentioned above have locally penetrated this ice after the deglaciation of this region of the Canadian Arctic. It is to be noted that the range in δ values within the ice bodies is comparable with the range observed for Pleistocene ice from the edge of the Barnes Ice Cap. This isotopic evidence for relic Pleistocene glacier ice in the western Canadian Arctic allows the reconstruction of the flow lines of the former Pleistocene ice sheet which covered the region.

In 1988, French and Harry discussed the nature and origin of ground ice in the Sandhills Moraine on Southwest Banks Island (Fig. 3.29). On the basis of several lines of evidence and notably ice fabric analysis, they concluded that the ground ice in the Sandhills Moraine is best interpreted as buried glacial ice rather than segregated or injection-segregation ice.

The relatively high-debris content of buried glacier ice is probably an indication that it is basal ice, an idea developed by Hughes (1973), who has coined

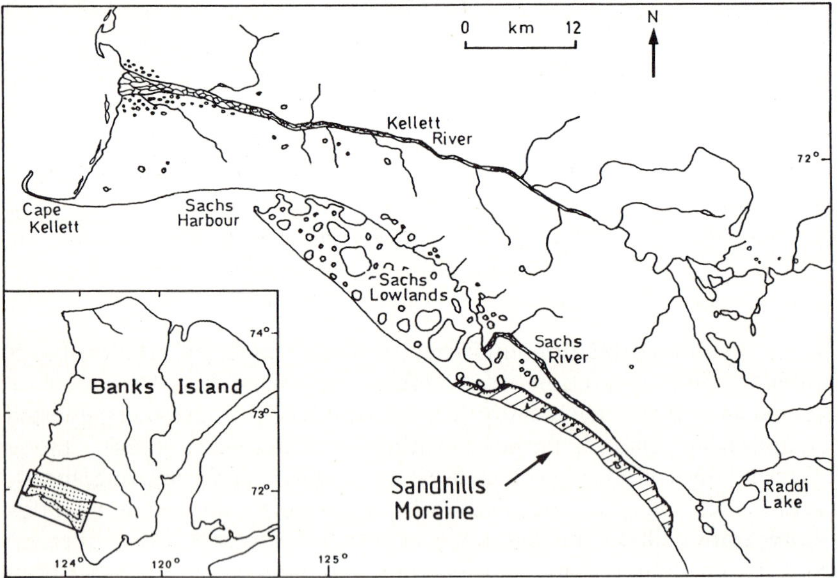

Fig. 3.29. Location map of Sandhills Moraine, southwest Banks Island. (After Fig. 1 in French and Harry 1988)

the term glacial permafrost for such a feature. Glacial permafrost will preserve its similarity to the debris-rich basal layer of an ice sheet for longer at depth. Near the surface, interaction with the active layer will result in ice wedging, frost heaving, patterned ground formation and other surface or near-surface phenomena that obliterate most of the initial character of the debris-rich glacial ice. At depth, however, many characteristics of the glacial basal zone should be preserved. When such evidence exists, it indicates the existence of a wet-freezing zone at the base of the former ice sheet (Hughes 1973). Indeed, for a thick basal ice sequence to be constructed, a situation at the base of the ice sheet must exist such that the amount of heat provided at the glacier sole is sufficient to prevent perennial freezing. In such a case, the latent heat released by the freezing of meltwater is sufficient to maintain the glacier sole at the melting point and to allow ice sheet sliding. Glacial permafrost can thus help to reconstruct the thermal regime of a Pleistocene ice sheet at its maximum development. This is an essential step for modelling the surface profile of such an ice sheet and for the computation of the ice volume, a question of importance in palaeoclimatic reconstructions. Distribution of glacial permafrost is limited to the region of continuous permafrost inside the zone uncovered by a Pleistocene ice sheet. As such, Northern Canada and Siberia both represent regions where potential glacial permafrost exists. Within this context, Kaplyanskaya and Tarnogradsky (1976) found relict glacier ice in the north of western Siberia and Solomatin (1981) has pursued the question further in the same region. A major problem is that it is difficult in some instances to distinguish between glacial permafrost and debris-rich segregation ice. The preservation of glacial permafrost after retreat and disappearance of the ice sheet from the region is due to the fact that surface melting creates an insulating cover of till on top of the ice layer. It is thus not surprising that no significant difference in grain size or mineralogy can be observed between the till above the ice body and the particles inside the ice body itself.

Ice composition of buried glacier ice can be used to derive valuable information for reconstructing the flow pattern and configuration of ice sheets during the Last Glacial Maximum.

4 The Basal Zone of Ice Caps and Ice Sheets

4.1 Thermal Conditions at the Glacier Sole

As indicated by Hooke (1977), the steady-state temperature distribution in a polar ice sheet is determined by the following parameters: the rate of influx of geothermal heat at the base, the temperature near the surface (usually the temperature at a depth of 10 m to avoid seasonal variations) and the vertical and horizontal components of the velocity at every point in the glacier. The term steady state is used here to describe a distribution of temperature in space that does not change with time. For a stagnant ice mass everywhere below the pressure melting point, the temperature gradient is determined by the geothermal heat flux. If, for example, the ice mass is 1000 m thick and the temperature gradient is 0.02 °C/m, a surface temperature of −21 °C will induce a basal temperature of −1 °C. If now the surface temperature is −10 °C, the mean temperature gradient through the ice mass must be adjusted such that the basal temperature does not exceed the pressure melting point (i.e. 0.01 °C/m in the ice sheet cited above). Thus, part of the geothermal heat cannot be conducted upward through the glacier and is consumed by melting ice at the sole. If half of the geothermal heat is used in melting ice at the glacier base, then about 1.8 mm of ice would melt per year. Water produced at one place at the base of the ice sheet is then able to move to another place where the temperature regime may be different and where the water might refreeze. In order to refreeze this water, the temperature gradient at the new location must be greater than some threshold value (0.02 °C/m in the example above).

The effect of vertical velocity must also be considered. The vertical velocity vector is directed downwards in the accumulation zone and is, at the ice surface, equal to the accumulation rate. Similarly, in the ablation zone, the vertical velocity vector at the surface is directed upwards and is equal to the ablation rate. At the bed the vertical velocity vector is nearly zero, except for small fluctuations resulting from melting or freezing. The rate of decrease in the vertical component of velocity is higher near the surface than near the bed. The downward-moving elements of ice are colder than the part of the glacier to which they are moving, with the result that at each level in the glacier heat is used to warm the downward-moving ice. This heat is supplied by the total heat flux moving upwards and the temperature gradient necessary to conduct that heat is correspondingly reduced. As vertical velocity is upwards in the ablation zone, the ice must be cooled as it rises and heat thus added to the geothermal flux, thereby increasing the temperature gradient. Hooke (1977) gives calculat-

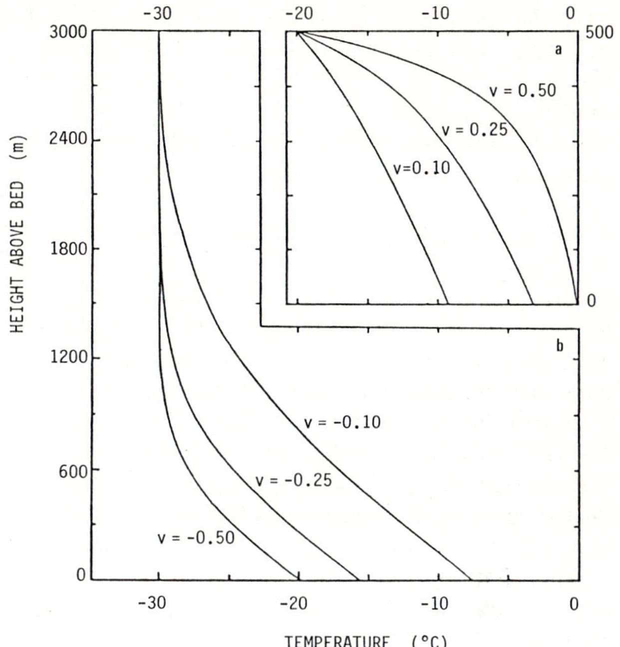

Fig. 4.1. Illustrative temperature profiles. v = vertical velocity at glacier surface and is positive upward. **a** Ablation zone. Vertical component of velocity is upward. **b** Accumulation zone. Vertical component of velocity is downward. Horizontal conduction and advection and internal heat generation are ignored in all cases. Vertical velocity is assumed to decrease linearly with depth. (After Fig. 3 in Hooke 1977)

ed temperature profiles for realistic combinations of ice thickness and vertical velocity, reproduced in Fig. 4.1. Stipulated boundary conditions – same surface temperature and same basal gradient – and the same ice thickness are used for the three situations depicted in Fig. 4.1. It is to be noted that the basal temperature increases markedly with increasing vertical velocity in the ablation zone and this may be one of the most common causes of basal melting. Temperature profiles in the accumulation zone are displayed in Fig. 4.1b. In contrast to the situation in the ablation zone, an increase in vertical velocity here results in a decrease in basal temperature. The basal temperature does not necessarily increase if the ice thickness is high in the accumulation zone. If the vertical velocity is high, the basal temperature may decrease because the new ice surface is colder, since it is at a higher elevation.

Considering the effect of horizontal velocity, the horizontal transport of heat by the outward movement of an ice column can be termed advection, from a term used by meteorologists. Here, advection is used to refer to the transport of heat by the outward mass movement of ice. Robin (1983a) indicates that the faster the mass movement, the more rapidly will basal heat be carried away by the horizontal motion of the ice sheet, such that less heat is available to flow to the upper layers of the ice sheet. Because the velocity decreases with depth, there is shear deformation in the horizontal direction which produces heat. The rate of deformation increases with temperature so that, in a polar glacier, the shear is particularly concentrated close to the bed. The frictional heat released is simply added to the geothermal heat with the result that the boundary condition is a higher temperature gradient at the bed. Hooke (1977) indicates that, for a typical basal shear stress of 0.5 bar, the frictional heat resulting from a velocity of 24 m/year is as great as the geothermal heat. Thus, the temperature gradient at the bed would have to be double that necessary to remove the geothermal heat flux alone. Evidently, the effect of frictional heating on ice temperature distribution is important. At an ice divide, the horizontal velocity is zero so that the frictional heat term is considered to be negligible. Frictional heating changes in line with horizontal velocity, which, in general, increases from the ice divide to the equilibrium line and then decreases downglacier to the margin.

The temperature gradient just above the glacier sole determines how much heat can enter the glacier per unit time. Three situations occur at the interface. In the first case, more heat – from the geothermal flux and from frictional heating – is provided at the glacier sole than can be conducted upwards through the glacier. Net melting occurs at the base; the glacier sole is at the melting point and the glacier is thus likely to slide over its bed. In the second case, the heat provided at the glacier sole per unit time is about the same as the amount which can be conducted through the glacier. Here there is a balance between melting and freezing, the glacier sole is at the melting point and the glacier is likely to slide over its bed. Pressure melting takes place on the up-glacier flank of bed obstacles and refreezing of the resultant meltwater occurs on the down-glacier flank. Only a local thin basal ice layer due to refreezing at the base is likely to be present in this case. In the third case, the amount of heat provided at the glacier sole is insufficient to prevent generalized freezing. As indicated by Boulton (1972), two different situations can occur in such a case. First, meltwater produced in an adjacent zone flows toward the basal region considered and the latent heat released by the freezing of this meltwater is sufficient to maintain the glacier sole at the melting point. Both generalized basal freezing and sliding of the glacier over its bed take place, a situation in which thick basal ice sequences can accumulate. This net accretion of basal debris-charged ice is visualized in Fig. 4.2. In the second situation, the glacier is frozen to its bed and the strong adhesion between ice and the substrata at temperatures below the melting point reduces sliding considerably. The forward movement of the ice sheet is by internal flow alone. Different thermal zones can thus be considered at the base of an ice sheet which play a paramount role in glacier dynamics and subglacial erosion. The areal distribution of these zones is likely to change in the course of time as a result of climatic changes. Present-day basal temperatures are below freezing for the major part of the East Antarctic Ice Sheet and for parts of the West Antarctic Ice Sheet, although a combination of thick ice and low surface elevations and consequently higher annual mean surface temperatures indicate that the bed may be at the

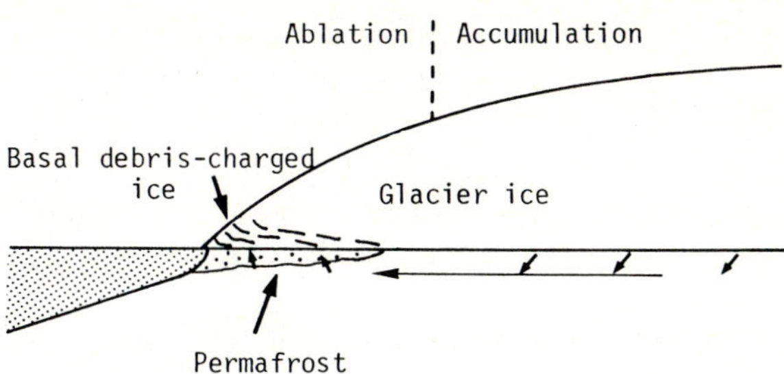

Fig. 4.2. Situation in which basal freezing and sliding of a glacier over its bed take place. *Fine arrows* indicate meltwater movement. Thick basal ice sequences can be produced. (After Fig. 1 in Boulton 1972)

pressure melting point over wide areas. The same kind of calculation suggests that basal temperatures beneath Central Greenland are also below the freezing point. Theoretical models predict zones of contrasting basal thermal conditions beneath ice sheets. In the central zone, basal temperatures are either below or at the pressure melting point, depending on the distribution functions used for the different model parameters. Of great interest is the fact that in all models a zone of basal melting exists near the margin, followed in places by a zone still closer to the margin where the glacier is again frozen to its bed as the ice thins rapidly. These different thermal zones can be combined in different ways. Figure 4.3 is a reconstruction from Sugden (1977) of the zones of contrasting basal thermal regime beneath the Laurentide Ice Sheet at its maximum.

4.2 The Effective Bed

Examination of large areas recently uncovered by Pleistocene glaciers leads to the perception that parts of ice sheets are underlain by unconsolidated sediments. In such areas, the rheological properties of the sediment must, to a certain point, control the dynamic behaviour of the ice sheet. The upper and lower interfaces of these unconsolidated sediments, as well as any internal surface, must be considered as a potential effective bed for an ice sheet. The concept of the effective bed is based on a surface where the forward motion is nil. A sliding interface is the most obvious effective bed, but this is not the only case. If deformation occurs within the subglacial sediments, then the surface of zero velocity may be within the unconsolidated material rather than, in the more classical case, at its upper margin.

The position of the melting point isotherm plays a great role in this context. Let us consider two different situations. On the one hand, the melting point isotherm is within the unconsolidated sediments and above this, they exist as frozen ground, often no longer unconsolidated. On the other hand, the melting point isotherm is at the limit between the subglacial sediments and bed-

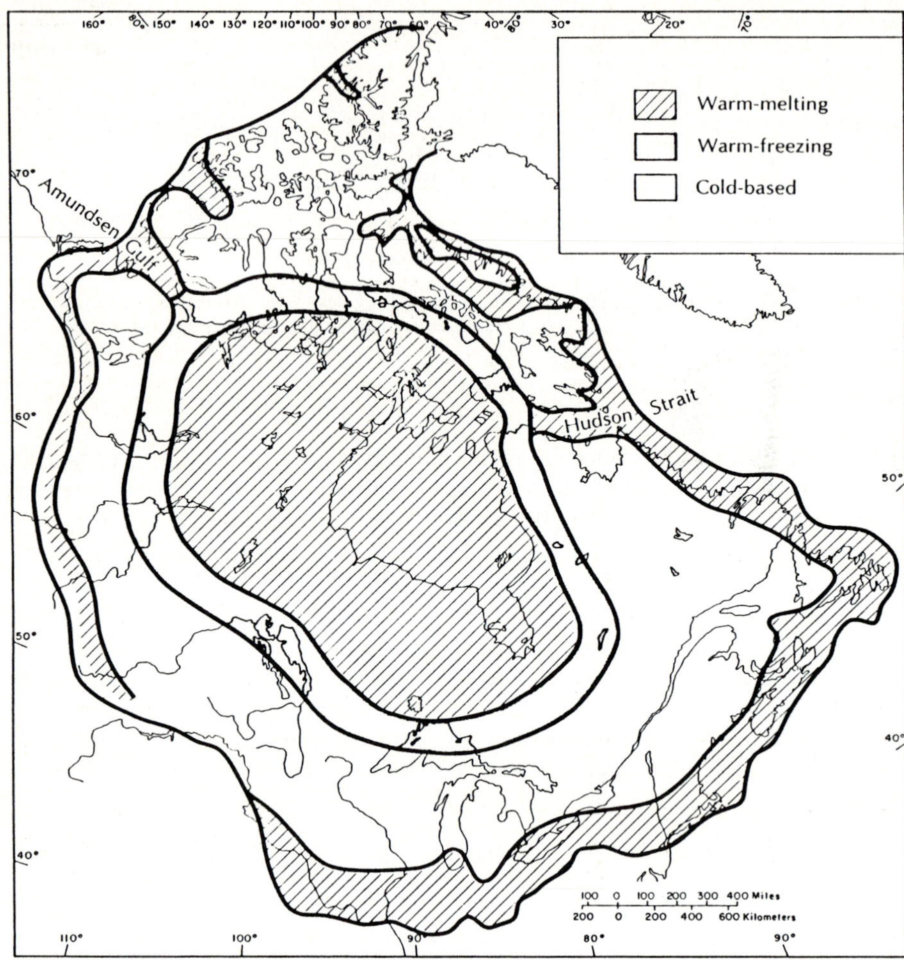

Fig. 4.3. Zones of contrasting basal thermal regime beneath the Laurentide ice sheet at its maximum. The warm-melting basal regime is characterized by melting at the pressure-melting point at the base, the warm-freezing regime by freezing at the pressure-melting point and the cold-based regime by a temperature below the pressure-melting point. The outermost zone where the glacier is again frozen to its bed is not represented. (Sugden 1977, Fig. 11)

rock, such that all the sediment is now frozen. When basal sliding occurs at the lower limit of the frozen ground in these two cases, the permafrost layer becomes incorporated en masse into the ice sheet. If, following Hughes (1973), this contact is a melting interface, then the permafrost layer will be eroded from below and the eroded material will be washed out towards the edge of the ice sheet. If this interface is a freezing interface, the lower part of the unconsolidated material or the bedrock, depending on the case envisaged, will be eroded from above and the eroded material will be frozen onto the base of the permafrost layer.

Let us now suppose that the melting point isotherm coincides with the upper part of the unconsolidated sediment, i.e. with the glacier sole. The base of the ice sheet may be a melting or a freezing surface. In general, the net freezing zone lies downglacier of a net melting zone. If this is the case, then water which forms a thin sheet along the bed and flows outwards towards the ice sheet margin, due to the decrease in hydrostatic pressure, will refreeze closer to the margin. Basal erosion occurs in the melting zone and the eroded debris is washed into the freezing zone, where it is incorporated into the ice sheet, provided meltwater does not percolate through the underlying unconsolidated sediment. Weertman (1966) considers this problem of percolation and states that sheet flow in a basal water layer of thickness h_w is more important than percolation flow in an unconsolidated sediment of thickness h_s when $h_s < h_w^3/12k$, where k is the sediment permeability. With a melting point isotherm coinciding with the upper surface of the unconsolidated material, the subglacial sediment can be saturated with water and apt to deformation and this deformation is more likely to occur under high pore water pressures. Figure 4.4, from Boulton (1979), shows till deformation at the base of the Breidamerkurjökull in Iceland, which accounts for 90% of the forward movement of the glacier sole. Ice stream B displacement in West Antarctica also seems to be largely controlled by this process.

Further insight into the study of frozen ground displacement at the base of an ice sheet can be gained by considering the creep behaviour of permafrost. Frozen soils exhibit substantial deformation under sustained loading and an important influence here is the unfrozen water content of the frozen ground. Figure 4.5 shows the amount of unfrozen water for different materials as a function of the temperature below freezing. If the frozen soil consists of fine particles, unfrozen water remains important at temperatures significantly lower than the freezing point, the amount of liquid water diminishing with decreasing temperature. In a review of published data, Anderson and Morgenstern (1973) consider that the unfrozen water contents of most frozen soils are conveniently represented by a simple power curve:

$$w = m\,\Theta^n ,$$

where w is the water content, m and n are characteristic soil parameters and Θ is temperature, expressed as a positive number in °C below freezing. Unfrozen water is in the form of thin films at the surface of soil particles while the ice is located in the pores of the soil, if supersaturation is not reached. This thin layer of liquid water which may exist down to low temperatures, has a thickness of a few angströms (between 5 to 10 Å). Nearly perfect slip can occur at ice-particle and particle-particle interfaces if this layer is continuous enough. This can then lead to a significant decrease in the effective viscosity of the bulk material. The application of pressure at constant temperature increases the unfrozen water content of a frozen soil. Clear-cut evidence that the thickness of the unfrozen water interface increases with increasing pressure is provided by the observations that the electric conductance of a frozen clay-water mixture increases dramatically with pressure.

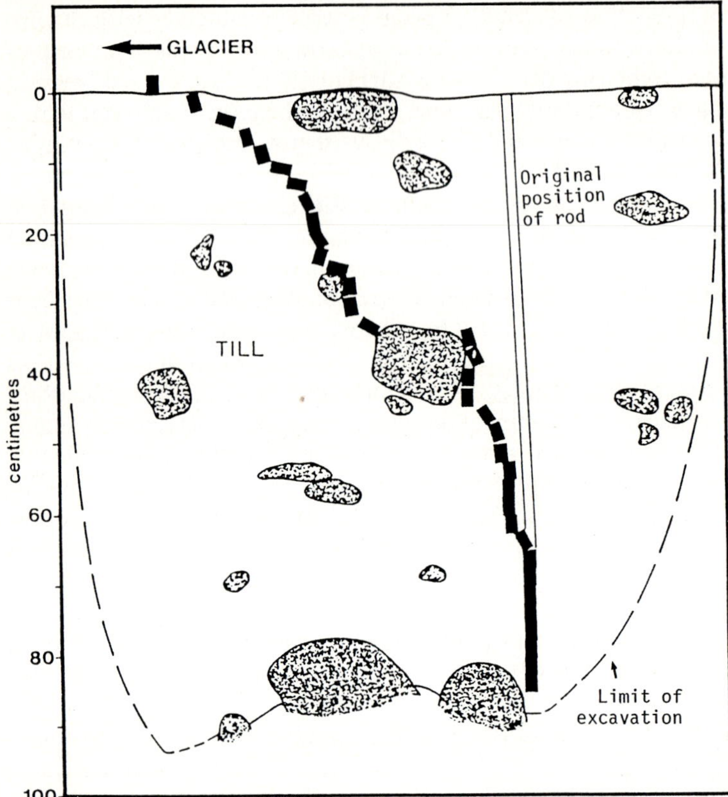

Fig. 4.4. Deformation of till beneath Breidamerkurjökull (Iceland); the position of individual annuli at one spot, 244 h after their original emplacement in the position shown is indicated. Boulders clearly strongly influence the pattern of deformation. The discharge of till at this site is 1122 cm^3/cm in 244 h. (Boulton 1979, Fig. 7)

Creep tests on permafrost soils show that this material, like ice, exhibits steady-state creep behaviour, which can be described by the equation

$$\varepsilon = B\, \sigma^n ,$$

where ε is the strain rate, σ is the stress, n is a viscoplastic parameter and B depends on temperature and material properties. The effective viscosity π can be defined as σ/ε. Figure 4.6 from Hughes (1973) shows π versus σ for pure ice and for two representative permafrost soils: clay permafrost and sandy loam permafrost. At stresses above 40 bar, these permafrost soils creep more readily than pure polycrystalline ice. However, under basal shear stresses commonly encountered at the base of an ice sheet (≈ 1 bar) ice creeps more easily than these permafrost soils. The study of Thompson and Sayles (1972) on the in situ creep of naturally frozen silts and gravels in the permafrost tunnel of USA-CRREL at Fairbanks (Alaska) has confirmed that the stress-strain relationship

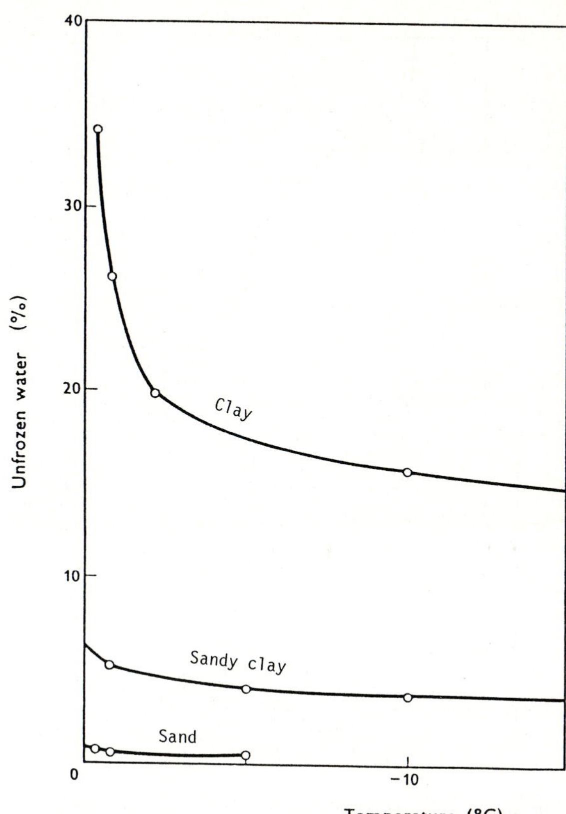

Fig. 4.5. Unfrozen water content of frozen soils against temperature. (After Fig. 1 in Tsytovich 1957)

in frozen materials during steady-state creep is described well by the flow law given above. The best fit to the data has been obtained with a value for n of 3 or 4 and, at temperatures around −2 °C, supersaturated silt deformed almost two orders of magnitude faster than non-saturated gravels. Ice-rich silt deformed 3.3 times faster in the field than in the laboratory, a rate which appears to be faster than that normal for pure ice alone. Haeberli (1985) has found that the creep rate of perennially frozen, ice-rich, rock glacier sediments, at −1 °C is smaller than the creep rate of pure ice by at least one to two orders of magnitude. While Haeberli's observations were of active rock glaciers in discontinuous Alpine permafrost, studies of rock glaciers in the continuous permafrost zone could furnish important information about the creep behaviour of ice-rich permafrost under different temperature conditions. There seems thus to be no single answer to the question of how an ice-laden sediment behaves rheologically as compared with pure glacier ice. Recent observations of the deformation of ice-laden drift in a tunnel at the base of a cold glacier in China by Echelmeyer and Zhongxiang (1987), reported in Section 1.3, add to the complexity. They have observed that deformation of the subglacial ice-laden drift is responsible for a major part of the overall surface motion of the glacier.

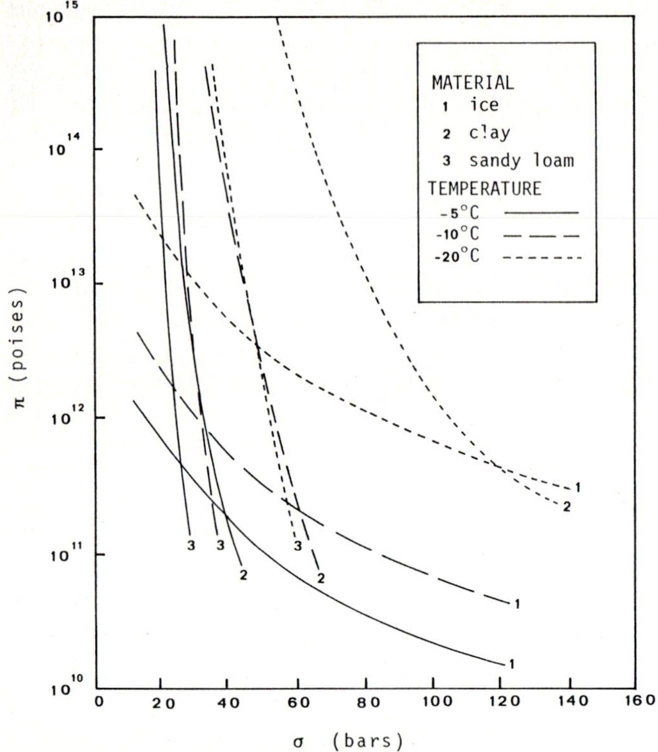

Fig. 4.6. A comparison of the stress σ and temperature T variation of effective viscosity π for polycrystalline ice, clay permafrost, and sandy loam permafrost. (Hughes 1973, Fig. 1)

Apart from the role played by thin films of unfrozen water, the authors consider those properties of granular materials relevant to the deformation of the ice-laden drift. The drift can be considered as a dense suspension of particles in a viscous fluid and, since it contains a large volume concentration of rock particles, ranging in size from clay to boulders, granulo-viscous effects are important in its deformation. Intergranular friction occurs both between soil grains and between ice and soil grains. This internal friction may allow failure to occur within the drift and, possibly, leads to the development of shear planes.

Boulton (1979) has envisaged the situation of a cold Antarctic glacier having overridden a series of sand dunes cemented by ice, the dunes having originally formed from a mixture of blown sand and blown snow. The ice layers in them represent metamorphosed snow. The frozen sediment bed of the glacier has deformed both by fracture and flow and masses of sediment have been plucked from the bed without disaggregation. There are two reasons for this plucking. First, the adhesive bonds between glacier ice and frozen bed sediments are strong and the frozen sediments are subjected to a high tractive force. This force is probably enhanced by a greater roughness at small scale due to spheroidal grains. Second, the existence of ice strata, of zones of low ice saturation, of intergranular voids and of open voids, tend to facilitate the process. Freezing of subglacial water is not likely to occur in this case as there is no free water. Therefore, an incorporation mechanism by freezing-on similar to the one envisaged by Weertman (1961), is certainly not operating at the base of this Antarctic glacier.

The same rapid deformation as that occurring beneath the glacier sole of a temperate glacier like Breidamerkurjökull, Iceland (Boulton 1979) can also be present under a cold glacier. If the unlithified sediment is thick and only its upper part is frozen, large-scale subglacial deformation beneath the frozen-unfrozen sediment interface is possible where high pore water pressures produce low sediment strength. The shear strength τ^* of unlithified sediment can be expressed by the Coulomb law:

$$\tau^* = c + (P - P_w) \tan \phi \ ,$$

where c is the till cohesion, ϕ is its angle of internal friction, P is the ice pressure and P_w is the pore water pressure. A high pore water pressure will thus reduce considerably the shear strength of unlithified sediments, favouring their subglacial deformation. Boulton and Hindmarsh (1987) have established the flow law for a deforming sediment beneath Breidamerkurjökull for relatively simple stress states. The strain rate ε is given by

$$\varepsilon = K\ (\tau - \tau^*)^a / N^b \ ,$$

with $\tau > \tau^*$ where τ is the shear stress, τ^* is the sediment strength, N is the effective stress or $(P - P_w)$ and a, b, and K are constants. A reduced sediment strength leads to an increased strain rate. Folding of unlithified sediments, as observed subglacially, is the consequence of such deformation. The folds have flat-lying axial planes with synclinal axes pointing upglacier and highly attenuated anticlinal fold closures.

4.3 The Basal Zone in Ice Cores

The properties of basal ice obtained from ice cores can shed some light on the dynamics at the base of an ice sheet or an ice cap. However, not many cores have been recovered from the basal zone of either the Greenland or the Antarctic ice sheets. The Camp Century ice core, drilled in North-West Greenland, is certainly a classical one and will thus be described first. Deep polar ice core records are listed in Table 4.1, from Lorius (pers. commun.). It can be seen that both the Camp Century and the Dye 3 deep cores reach basal ice, while, in Antarctica, only the Byrd core was drilled to the bedrock.

Camp Century is located in North-West Greenland, approximately 200 km from the west coast and 500 km from the present ice divide. It is in the accumulation zone of the ice sheet, in an area where there is some summer melt. The basal zone, characterized by visible silt bands and pebbles, comprises the bot-

Table 4.1. Deep polar ice core records. (Courtesy Dr. C. Lorius, Laboratoire de Glaciologie du CNRS, Grenoble)

	Drillings		Current conditions			
	Depth (m)	Year	Ice thickness	Elevation (m)	Mean temperature (°C)	Mean accumulation (m^2/a)
Greenland						
Camp Century 77°10′N 61°08′W	1387	1966	1387	1885	−24	32
Dye 3 65°11′N 43°50′W	2037	1981	2037	2480	−19.6	50
Antarctica						
Byrd 79°59′S 120°01′W	2163	1968	2163	1530	−28	16
Dome C 74°40′S 124°10′E	905	1978	3400	3240	−53	3.4
Vostok 78°28′S 106°48′E	950	1974	3700	3490	−55.5	2.3
	2083	1982				
	2202	1985				

tom 16 m of the core, which has a length of 1387 m. The temperature at the base of the ice sheet at the location of the core is −13 °C and the core spans an estimated time period of about 125000 years (Dansgaard et al. 1971). This estimation is based on a flow model assuming that meltwater has never existed at the base of the ice sheet. As indicated by Herron and Langway (1979) the mechanisms by which the basal debris was incorporated are a critical element in estimating the age of the ice core. If the freezing-on mechanism proposed by Weertman (1961) has occurred, then the age determination may be an underestimation.

The investigations carried out by Herron and Langway (1979) can be divided into several component studies:

a) the stratigraphy of the basal ice,
b) the size, concentration and composition of the embedded debris,
c) the ice textures and fabrics, and
d) the volume and composition of gas inclusions.

The basal 16 m of the Camp Century ice core contains over 300 alternating bands of clear and debris-laden ice. On the basis of this banded structure and the distinct boundaries between the debris bands, it is possible to reject any type of dispersion or diffusion along temperature, pressure or concentration gradients as the primary mechanism for the emplacement of the basal debris. A general increase in clay-size material with increasing distance from the bottom and a corresponding decrease in sand-size material has been observed. There are also many particle aggregates in the basal ice, and this may result from the incorporation of unconsolidated sediments. Moreover, unconsolidated, frozen till-like material is present directly beneath the debris-laden ice. The average debris concentration is 0.24% by weight for the entire 16-m thick basal zone as compared with 0.001% by weight in ice above the debris-laden zone. This high debris concentration supports a subglacial origin for the debris. X-ray analysis of the particles indicates a predominance of quartz with albite, orthoclase, kaolinite and hornblende. The similarity of all the X-ray patterns indicates that there is no significant mineralogical change over the 16-m debris-laden profile. Air bubbles and disc-shaped inclusions are observed parallel to the debris bands, yet there is no bubble stratification. The ice crystals are quite small, with an average diameter of 0.8 mm in the lower 10 m of the core. This could be due to the existence of differential shearing. In the bottom 10 m, the ice exhibits a highly preferred vertical orientation of crystal optic axes. Above, the degree of preferred orientation diminishes and the average ice crystal diameter reaches about 2 mm. The maximum shear stress and high deformation therefore seems to be concentrated in the lowest 10 m of the ice sheet. The average gas concentration is 5 cm^3/100 g of ice, 50% lower than in the rest of the Camp Century core. This average total gas content is about midway between the expected values of near zero for a simple freezing model and close to 10 for the overlying glacier ice. Herron and Langway (1979) consider that bubble-free debris-laden ice was originally incorporated at the base of the ice sheet by a simple freeze-on mechanism and that soluble gases then diffused through the debris-laden ice at a temperature close to the pressure-melting point. The gas concentration profile shown in Fig. 4.7 might result from such a process. If gaseous diffusion has occurred, then it should be reflected in the gas composition. Indeed the most soluble gas component, argon, is most enriched in samples with the least amount of air. This is coupled with a depletion of argon in the upper zone where diffusion originated. This pattern is considered to be the consequence of the combined effect of the high solubility of argon in liquid water and its rapid diffusion relative to N_2. Air may have diffused from the overlying glacier ice with ordinary gas content and composition into the basal layer of refrozen meltwater with very low gas content. The gas analysis thus provides evidence which supports the freezing-on mechanism for the origin of the basal ice.

During the latter stages of the drilling of the 2163-m deep hole at Byrd Station in West Antarctica, liquid water was encountered at the ice-bed interface (Gow et al. 1968). This is clear evidence that the bottom of the ice sheet at this location is at the pressure-melting point. The subglacial material is considered to be composed of unconsolidated sediments. The debris-rich basal zone containing predominantly clay-, sand- and pebble-size particles extends for 4.83 m above the bed. Larger rock pieces have also been recovered. most pebbles are sedimentary aggregates of clay and sand, held together by interstitial ice and are best described as "mud clots". Dirt-free ice also exists in the basal zone at Byrd Station. Gow et al. (1979) consider the entrapped air content along with stable isotope analysis in order to understand the mechanism by which basal debris is incorporated into the ice sheet. Only in the bottom 4.83 m of ice core is the air content of the ice observed to diminish to practically zero. Values

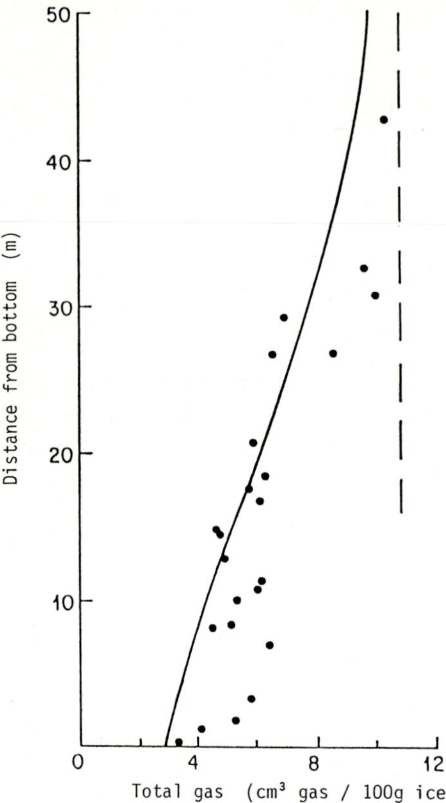

Fig. 4.7. Gas-concentration profile in the basal ice layers of the Camp Century core. (Herron and Langway 1979, Fig. 5)

measured are around $0.2\,cm^3$ of air/100 g of ice, compared with $10-12.5\,cm^3/100\,g$ in glacier ice above the basal zone. The transition from air-rich glacier ice to ice virtually devoid of air coincides precisely with the first appearance of stratified debris in the cores. This adds weight to the idea that the basal sequence originated from the refreezing of meltwater at a sufficiently low freezing rate to allow the complete rejection of air dissolved in the water. Concerning the oxygen isotopic analysis of the ice in the basal zone of the Byrd core, the mean $\delta^{18}O$ value of the bottom ice samples is similar to that obtained from glacier ice samples above the basal zone but variations in $\delta^{18}O$ values among closely spaced samples in the basal 4.83 m exceed those observed in the overlying glacier ice. A sample of basal meltwater has a $\delta^{18}O$ value about 3‰ lower than basal ice and this reinforces the case for bottom freezing, since the maximum enrichment of ice derived from the freezing of water is of this order. Figure 4.8 gives the entrapped-gas and $\delta^{18}O$ measurements for ice in the bottom 10 m of the Byrd Station core. There appears to be no systematic relationship between the $\delta^{18}O$ values and the debris concentrations in the ice. The ice crystals from the base of the Byrd core are about two orders of magnitude larger than those at Camp Century and the fabric orientation is much

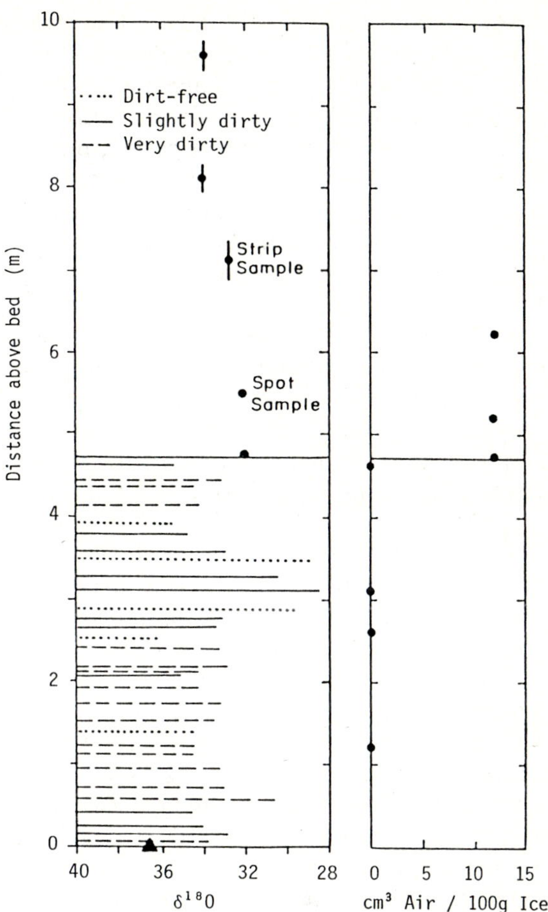

Fig. 4.8. $\delta^{18}O$ and entrapped-gas measurements in the bottom 10 m of ice at Byrd Station, Antarctica. Transition between true glacial and basally accreted ice occurs at 4.83 m above the bed. Note that dimensions of vertical strip samples are denoted by length of line through data points; spot samples measured 4–5 cm long; basal melt water is represented by a triangle. (Gow et al. 1979, Fig. 5)

weaker than that observed in the overlying ice (Gow 1970). The striking difference between the two cores reflects the difference in thermal conditions: basal temperature is at the pressure-melting point at Byrd while it is at $-13\,°C$ at Camp Century.

The Dye 3 deep ice core was drilled on the south Greenland Ice Sheet dome (Fig. 1.15 b). The ice core is 2035 m deep, the final 22 m consisting of silty ice, with an increasing concentration of pebbles downward. Between $y = 22$ m and $y = 50$ m, with y being the elevation above the bed, the Dye 3 oxygen isotopic record does not show significant δ variations in contrast with that part of the Camp Century core corresponding to same time interval. Dansgaard et al.

(1982) think that this may be due to flow complications, the drill probably penetrating the ice parallel to the isochrons. This suggests a folding of the deepest layers above the silty ice, perhaps due to the influence of nearby mountainous subglacial terrain. Dansgaard et al. (1982) have considered that isotopic values of the silty ice indicate a warm period, probably the last interglacial. Stauffer et al. (1985b) have found that its total gas content is only about two thirds of that above the silty layer, and CO_2 is considerably enriched, probably due to a high carbonate content since the gas was extracted with a melt extraction method so that carbonates contribute to the CO_2 content.

A number of ice cores, reaching the bed at a depth of about 100 m, have been retrieved in the coastal region of Terre Adélie in East Antarctica. Figure 4.9 gives the location of three drilling sites where the subglacial bed has been reached: respectively G_1, G_2 and C.A.R.O.L.I.N.E. In the three cores, the upper part consists of glacier ice of local origin, while an abrupt transition in δ values indicates the appearance of Pleistocene ice from the interior further down in the cores. In G_2 and C.A.R.O.L.I.N.E., the subglacial bed is reached above sea level after the penetration of a debris-rich basal zone, but in G_1 the ice-bed interface is below sea level. Figure 3.13 gives a cross-section in this coastal region (Nougier and Lorius 1969). It is clearly visible that the upward component of flow is responsible for bringing basal ice to the surface at the level of the moraine not far from Cape Prudhomme. In the G_1 core reaching the substratum below sea level, a layer of blue ice about 3.5 m thick is present beneath the debris-rich basal zone. Figure 3.14 from Lorius (1968) gives the δD profile, the mean crystal size profile, the gas content and the soluble impurity content in the G_1 core. High δD values and a high soluble impurity content (up to 20 mg Na per litre, not indicated on the figure because of the scale used) in the blue ice beneath the debris-rich basal zone undoubtedly indicate a marine origin for this ice.

In a paper published in 1986, Boulton and Spring have used the well-known Rayleigh model to interpret the oxygen-isotopic composition of basal ice from Byrd Station and to discuss its possible application to other oxygen-isotopic profiles from polar glaciers. This approach, however, is subject to several important limitations. While the fractionation at the water-ice interface is always given by the equilibrium fractionation coefficient (1.003 for ^{18}O), the amount of observed fractionation is actually dependent on the freezing rate. This is due to the fact that the water close to the ice interface is more or less depleted in heavy isotopes and the controlling factor is the ratio between the freezing rate and the diffusion coefficient of $H_2{}^{18}O$ in water, as seen in Chapter 2. As pointed out by Posey and Smith (1957), the consequence is that the ice is more or less enriched in ^{18}O. The apparent fractionation will be different from the true value. The isotopic range between the value at a certain percentage of freezing and the value of the initial water can therefore not be predicted unless diffusion phenomena and freezing kinetics are considered. It would only be fixed if water is always completely homogenized during freezing. This limiting case, probably never realized in nature, would give a Rayleigh distribution for an infinitely low freezing rate. Figure 4.10, from Boulton and

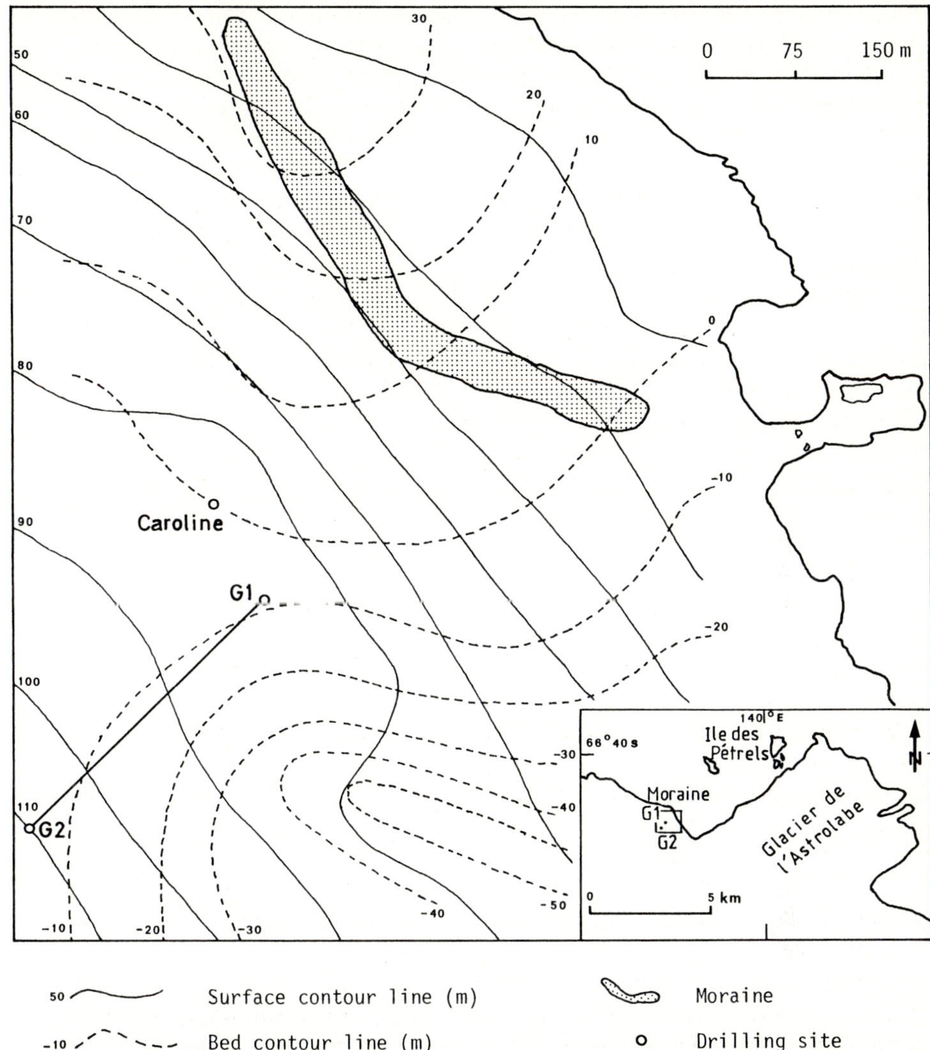

Fig. 4.9. Location map of the three drilling sites of the coastal region of Terre Adélie (Antarctica) where the bed was reached at a depth of about 100 m. (Courtesy Dr. J.R. Petit, Laboratoire de Glaciologie du CNRS, Grenoble)

Spring (1986), shows such a distribution. The significance of the isotopic composition of basal ice cannot be well understood by considering only the $^{18}O/^{16}O$ ratio. The concept of the freezing slope on a δD-$\delta^{18}O$ diagram developed by Jouzel and Souchez (1982) and Souchez and Jouzel (1984) must be considered and applied as in the next section. If an open system is considered, then another difficulty arises for the prediction of the $\delta^{18}O$ value of basal ice from the $\delta^{18}O$ of subglacial water as envisaged by Boulton and Spring (1986).

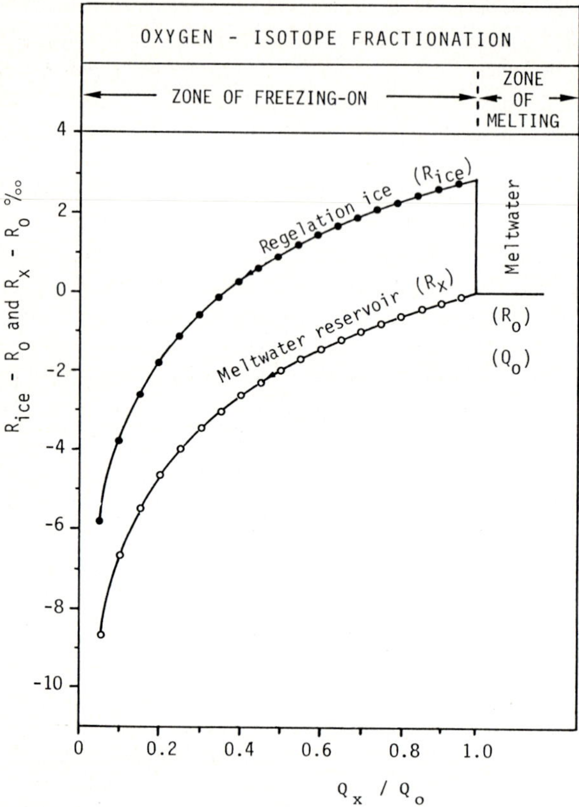

Fig. 4.10. Progressive oxygen isotope fractionation in subglacial water (R_x) and the overlying ice formed from it (R_{ice}) when an initial subglacial water discharge (Q_0) of isotopic composition (R_0) from a zone of melting is progressively frozen to the glacier sole in the succeeding zone so that subglacial discharge (Q_x/Q_0) diminishes. (Boulton and Spring 1986, Fig. 2)

Souchez and De Groote (1985) have shown that the freezing slope increases if initial water is mixed with isotopically more negative water during the course of freezing. Thus, the range of values for a single isotopic ratio between the value in ice at a certain percentage of freezing and the value of the initial water will be different if there is some mixing or if there is none. By a study of a single isotopic ratio, it is not possible to tell whether the isotopic profile in basal ice is due to fractionation by freezing in the absence of mixing.

4.4 Investigations in Marginal Areas

If ice caps and ice sheets do not reach the sea, then their marginal zones usually show a striking contrast between glacier ice, nearly devoid of visible particles, and basal ice, containing a rich sequence of subglacially derived debris layers. The margin of an ice sheet, ice cap or outlet glacier can be in the form of either an ice ramp or an ice cliff. Both situations commonly occur; they are the result of the interaction between flow rates and ablation rates. As the ice margin is not stable through time, recycling of debris can occur and this must be borne in mind during the analysis and interpretation of debris sequences.

The base of Aktineq Glacier, a Canadian Arctic outlet glacier flowing from an icefield located on Bylot Island, has been studied by Lorrain et al. (1981). Ice samples for isotopic study were recovered from the lower 3.5 m of a nearly vertical cliff about 14 m high on the left side of the glacier, 2.5 km from the snout. The lower 10 metres of this section can easily be differentiated from the bulk glacier ice above in that it contains a negligible amount of bubbly ice and numerous debris layers. The results of the analyses are plotted on a graph (Fig. 4.11, from Jouzel and Souchez 1982) with $\delta^{18}O$ as the abscissa and δD the ordinate. On this figure, two linear trends are observed. The points representing glacier ice samples lie on a straight line with a slope S = 7.8, as determined by linear regression. As indicated in Chapter 2, this trend is recognised in the literature as being a characteristic of rain and snow precipitation. The line on which the δD and $\delta^{18}O$ values are correlated is called the meteoric water line, described by the equation $\delta D = 8\ \delta^{18}O + 10$. The independent term, called the deuterium excess, can be different from 10. This slope of 8 is general-

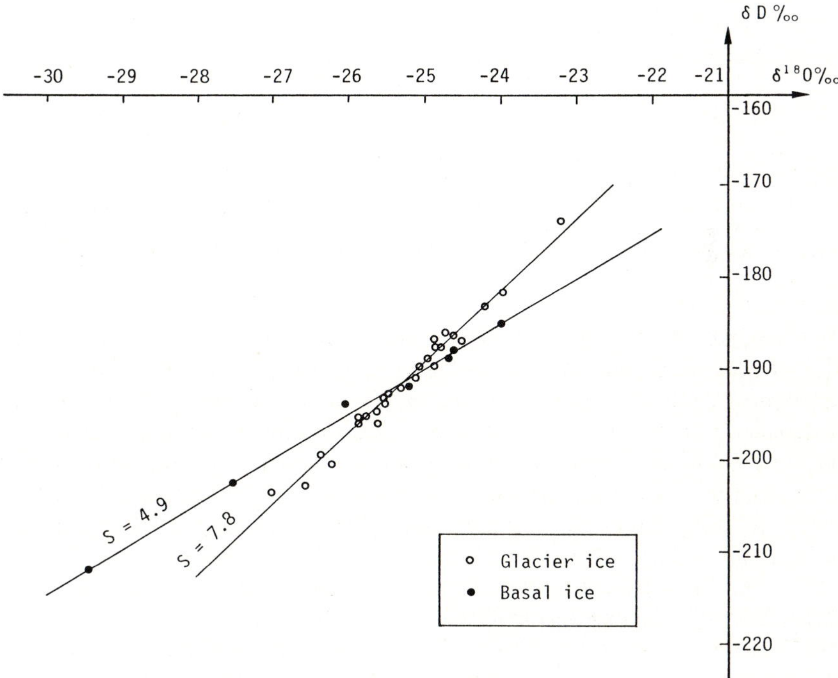

Fig. 4.11. The relationship between $\delta^{18}O$ and δD for samples of ice from Aktineq glacier, Bylot Island. (Jouzel and Souchez 1982, Fig. 1)

ly displayed in snow and glacier ice which has not been submitted to major isotopic changes since its formation. The other linear trend observed in Fig. 4.11 diverges from the line of slope 7.8 and is composed exclusively of basal ice samples located on a straight line with a slope of 4.9. In contrast to the meteoric water line, this slope represents the characteristic freezing slope for this area, as will now be demonstrated. The initial liquid at Aktineq Glacier most probably corresponds to the intersection between the two straight lines with slopes equal to 7.8 and 4.9. So, from Fig. 4.11, $\delta_i D = -192‰$ and $\delta_i{}^{18}O = -25.3‰$ for the initial water there. Using these values in the equation of the freezing slope

$$S = \frac{\alpha - 1}{\beta - 1} \cdot \frac{1000 + \delta_i D}{1000 + \delta_i{}^{18}O}$$

developed in Chapter 2, the slope calculated for the freezing process is 5.1. In the calculations of Jouzel and Souchez (1982), $\alpha = 1.0186$ for D and $\beta = 1.003$ for ^{18}O and the use of other values for the fractionation coefficients will not change the result appreciably. The slope of 4.9 obtained from the basal ice samples of Aktineq Glacier is in close agreement with that of 5.1 predicted by the model. Meltwater freezing as a mechanism for basal ice formation is thus strongly supported by a co-isotopic analysis of the ice. Gordon et al. (1988) comparing the isotopic composition of debris-rich and debris-poor basal ice of two glaciers in North Norway found a freezing slope developed on a δD-$\delta^{18}O$ diagram in the first case. These results, not only lend support to the works of Jouzel and Souchez (1982) and Souchez and Jouzel (1984), but show the association of debris entrainment with subglacial refreezing processes.

West Greenland provides a location where there is an extensive terrestrial ice sheet margin and where the crucial basal ice sequence can be examined in situ. Data have been collected from two sites, one at the snout of Russell Glacier near Søndre Strømfiord, representing a lobate margin, and one at the northern flank of the Jakobshavn Isbrae outlet glacier, representing a marginal zone of streaming flow (Fig. 1.15 b). Some glaciological characteristics of the two sites, including ice morphology, type and structures are shown in Fig. 4.12. The dominant feature in the extreme marginal zone in both cases is the presence of subparallel debris-rich ice layers up to 10 cm thick, dipping steeply upglacier. Clearly visible ice structures show the whole sequence is substantially folded, with the limbs of the folds more or less perpendicular to the ice flow.

Two basal ice facies can be distinguished in these glaciers (Sugden et al. 1987). On the one hand, a banded sequence of ice and debris layers is present at both sites. The debris layers occur both as discrete layers with only interstitial ice and as fine debris laminations, the two types often merging into each other laterally. The ice interbedded in the thinly laminated beds consists of layers of only one-crystal thickness. These ice crystals are up to 1 cm long, with no bubbles, although a few thicker layers may be bubbly. The debris consists mainly of sand and fine gravel, with considerable variation in the size distribution of individual samples (Fig. 4.13). Eighty percent of samples show a clear

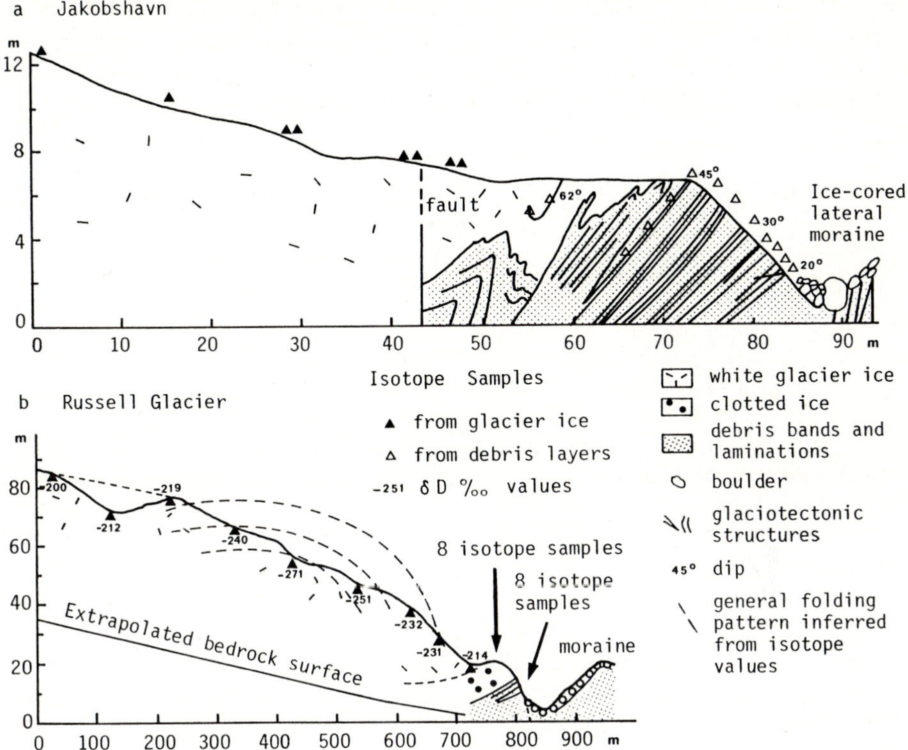

Fig. 4.12. The stratigraphy of the marginal ice at Jakobshavn Isbrae and the Russell Glacier, West Greenland. **a** The banded debris series at the northern margin of Jakobshavn Isbrae, showing folds and isotope sample sites. **b** Profile of the Russell Glacier showing three ice types and δ^{18}O values. The latter show that δ^{18}O values become less negative up the glacier, as would be expected in terms of normal glacier flow, but variations demonstrate the existence of a large discontinuity or fold. The approximate bedrock surface has been extrapolated from adjacent ice-free slopes. (Sugden et al. 1987, Fig. 2)

grain size frequency peak in the fine sand range (0.23 – 0.06 mm, $\approx 4\phi$) and in 34% of the samples there is a second peak at 6 – 20 mm ($\approx$ -3ϕ). As indicated in Chapter 2, this bimodal distribution is typical of glacial crushing, with one peak representing the mineral grain size of the gneissic parent bedrock and the other, rock fragments. None of the samples contains the fine silt or clay-sized particles typical of glacial abrasion. This probably reflects the periodic flushing of the ice-rock system by meltwater flowing at the base, illustrated by the presence of fine silts and clays in meltwater streams and marginal lakes in the area. This first ice type is equivalent to Lawson's stratified facies (Lawson 1979).

On the other hand, basal ice which contains a suspension of fine debris particles is composed of equi-dimensional ice crystals, 2 to 5 cm in size. The ice is essentially clear with only some crystal boundaries marked by pockets

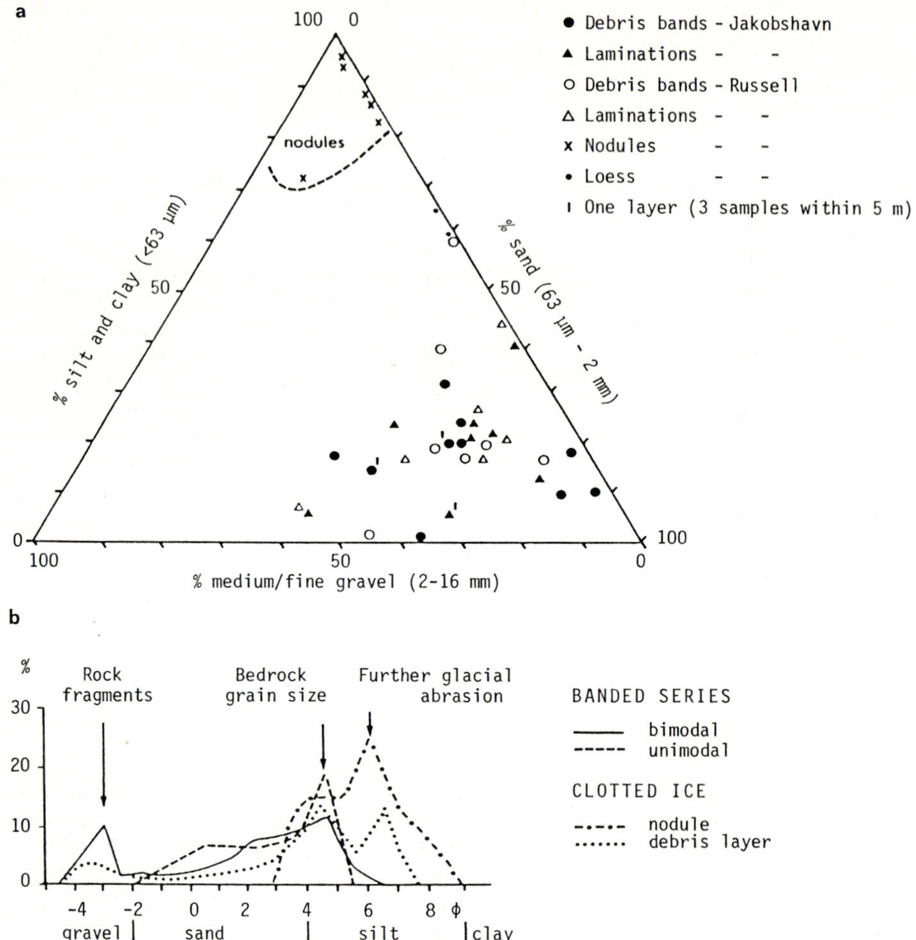

Fig. 4.13. Grain size of the rock debris in basal ice from Jakobshavn Isbrae and the Russell Glacier. **a** Triangular diagram showing the coarse nature and variability of debris in the banded series and the finer component in the nodules of the clotted ice. **b** The peaks in coarse silt and gravel fractions are typical of processes of subglacial crushing. The finer component due to abrasion is missing from the banded series but present in the nodules and debris concentrations in the "clotted" ice of Russell Glacier. (Sugden et al. 1987, Fig. 3)

of gas. It contains suspended lenticular nodules of debris up to a few cm in diameter, the size and concentration of which decrease towards the top of the sequence. In places, the nodules occur in planar clusters associated with small bubbles and occasional angular clasts. The debris in the nodules is different from that of the stratified facies in one major respect: it contains a significant finer component with 71–97% of the weight comprising silt and clay-sized particles. The fine debris is consistent with an origin from gneissic bedrock, as illustrated by the mineral assemblage (Table 4.2). The finer component due

Table 4.2. Constituent minerals of the clay-sized material in three clotted ice samples and their relative importance, Russell Glacier, West Greenland. (Sugden et al. 1987, Table 2)

Mica-trioct	3	4	4
Chlorite	2	3	2
Amphibole	3	2	2
Quartz	2	2	2
Pyroxene	2	2	2
Feldspar (plagioclase)	2	2	3
Feldspar (alkali)	1	1	1
1.0 – 1.4 nm interstratified	2	–	1

(1, trace; 2, <10%; 3, 10 – 40%; 4, >40%). The assemblage is typical of an origin from gneissic bedrock.

Table 4.3. Debris content in different types of basal ice, Russell Glacier and Jakobshavn Isbrae, West Greenland. (Sugden et al. 1987, Table 1)

	Sample number	Mean debris content by % weight
Banded series		
Debris bands Jakobshavn	10	55
Debris bands Russell	3	71
Debris laminations Jakobshavn	8	34
Debris laminations Russell	6	14
Clotted ice		
Nodule (7 × 5.5 × 4 cm)	1	76
Minimum (upper layers)	2	0.001
Maximum (lower layers)	3	4 – 8

to abrasion which is absent from the stratified facies is thus present in this "clotted" ice. The debris content of the two types of basal ice is given in Table 4.3. The clotted ice is probably equivalent to the dispersed facies of Lawson (1979).

The $\delta D/\delta^{18}O$ values from glacier ice at the top of the sequences lie along regression lines with a slope close to the value of eight. Samples of the stratified facies from the basal part of Jakobshavn Isbrae display a characteristic freezing slope on a δD-$\delta^{18}O$ diagram, as seen from Fig. 4.14. The freezing slope at the Jakobshavn Isbrae site is slightly higher than that predicted by the simple model: 6.3 instead of 5.5. This discrepancy is probably explained by mixing in the course of freezing with water which is more depleted in heavy isotopes. The clarity of the freezing slope for the Jakobshavn samples is due to the isotopic homogeneity of the parent water supplied for freezing by the overlying glacier ice. The range of values between the least negative sample on the freezing slope and the parent water is ≈4‰ in $\delta^{18}O$, too great to be accounted for by a single freezing event, which can only produce isotope enrich-

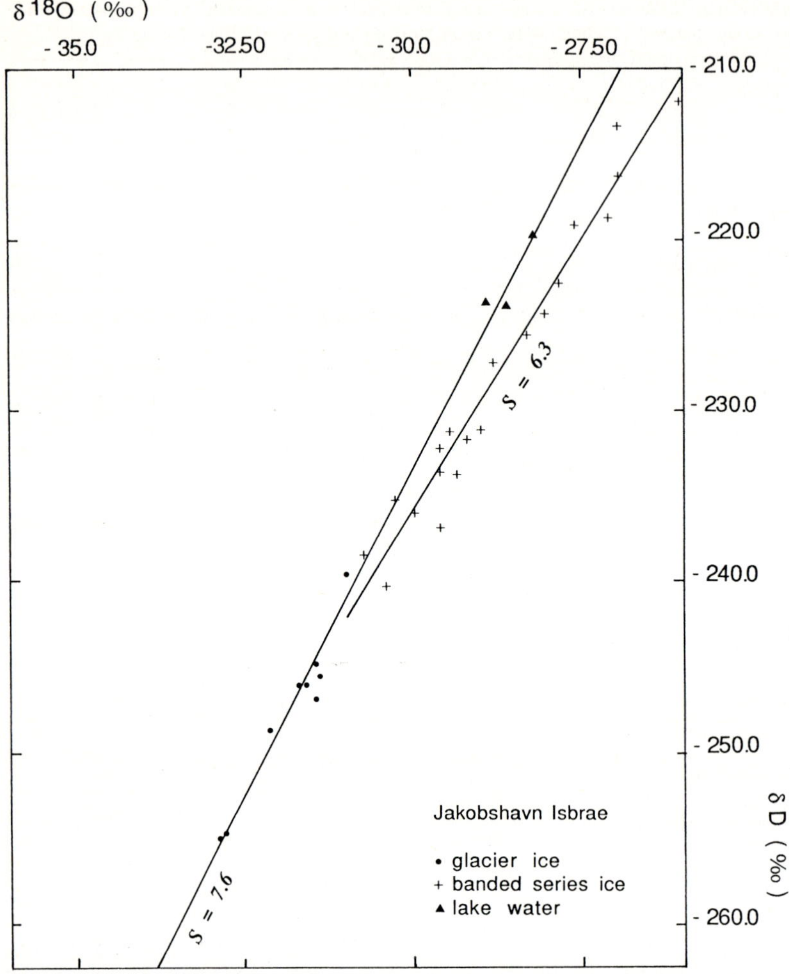

Fig. 4.14. δD-δ^{18}O diagram for ice samples from Jakobshavn Isbrae, Greenland. The slopes of the two regression lines are indicated by *S*. (Sugden et al. 1987, Fig. 4)

ment of up to 3‰ in δ^{18}O. This implies that ice in the debris layers has undergone multiple melting and freezing events.

In Russell Glacier, some ice samples from the dispersed facies are isotopically similar to the most negative glacier ice (Fig. 4.15). This glacier ice is probably of Pleistocene age or formed at the ice divide, since it is only at Crête in Central Greenland (approximately on the flow line ending at Russell Glacier) that present-day ice with a δ^{18}O value as low as −35‰ has been found (Robin 1983 b). Thus, it is likely that the clotted ice is derived from such isotopically light glacier ice. A regelation mechanism which occurs at the pressure melting point around bed protuberances may explain these isotopic charac-

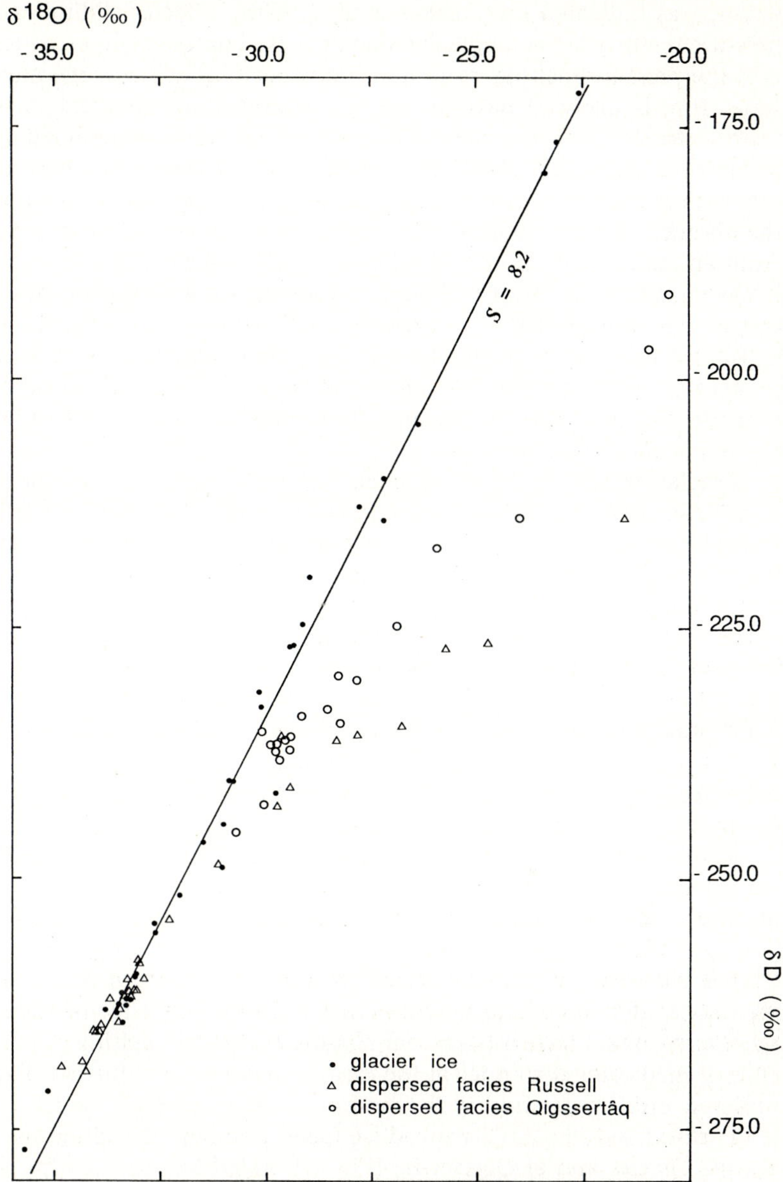

Fig. 4.15. δD-δ^{18}O diagram of samples from glacier ice and from the dispersed facies of basal ice at Russell Glacier and Qigssertâq, Greenland. (Souchez et al. 1990, Fig. 2)

teristics, as indicated by Souchez et al. (1988b). Glacier sliding can occur by pressure melting at the upglacier side of these bumps if the basal temperature is at the pressure-melting point and if a water film exists at the interface. The water film is likely to have an isotopic composition identical to that of the nearby glacier ice, as the melting of glacier ice is not accompanied by fractionation (Friedman et al. 1964). It is probably worth describing here a fractional melting experiment performed on glacier ice from Russell Glacier, supporting the absence of fractionation by melting. In a suitable hot-water bath, a slice from an ice core of about 125 ml was locally heated at the base. As soon as it was produced, 25 ml of meltwater was recovered with the aid of a polyethylene syringe and immediately press-filtered. All five successive 25-ml samples of filtered water were taken and their isotopic composition was analysed. The points representing these five successive samples are clustered on a δD-$\delta^{18}O$ diagram and as stated by Souchez et al. (1988b), this implies the absence of fractionation during melting.

Regelation ice can have the same isotopic composition as the bulk water in the film and so it retains the original isotopic character of the parent ice. Indeed, taking into account the developments given in Chapter 2, after an initial transient, a steady state in the isotopic composition of the ice at the level of the initial water δ value occurs at freezing rates higher than some critical value. Differential movement of the ice around a bump, as suggested by Boulton (1975), allows the incorporation of the regelation ice which has been formed into the basal part of the glacier and facilitates the dispersal of debris which was originally located at the interface. This dispersal can also be considered as the result of a kind of diffusion mechanism through an ice matrix undergoing simple shear deformation (Weertman 1968). The mud clots in the ice are probably derived from an unconsolidated drift at the glacier bed.

Regelation ice is produced in an interior zone where glacier ice at the pressure melting point with very negative δ values exists at the base. Clausen and Stauffer (1988) have found that the lower part of an ice core drilled in the marginal zone north-east from Jakobshavn is at the melting point temperature. Such a situation is probably frequent along the western and south-western margins of the Greenland Ice Sheet and only the very edge of the ice sheet in these areas could have a basal temperature below the melting point. The presence of meltwater within ice at the melting point and its further discharge are, however, unlikely to modify the isotopic composition of the ice.

Further study of the dispersed ice facies at other sites along the margin of Russell Glacier and at Qigssertâq (Fig. 1.15b) led to another type of distribution on a δD-$\delta^{18}O$ diagram. Indeed, some ice samples from this facies deviate significantly from the precipitation slope, without being aligned on a freezing slope (Fig. 4.15). A substantial enrichment in heavy isotopes appears for some of the samples from the dispersed facies. This enrichment is more important than the one resulting from a freezing process. It is also very different relatively for both isotopes. It is likely to be due to the interaction between hydroxyl-bearing minerals subjected to subglacial abrasion and water either at the glacier sole or in the ice where tiny amounts of meltwater were in contact with the mud clots (Souchez et al. 1990). As indicated above, ice from the dispersed facies is likely to have been at the melting point at depth. Water could thus have been present, accounting for the almost complete disappearance of gas bubbles. In the Jakobshavn area a freezing slope is formed by the data on a δD-$\delta^{18}O$ diagram for basal ice from the stratified facies; this is caused by the debris being sandy and devoid of fine particles resulting from glacial abrasion. By contrast, in basal ice from the dispersed facies at Russell Glacier and Qigssertâq, the presence of clay and silt aggregates with a substantial amount of hydroxyl-bearing minerals has a profound influence on the isotopic characteristics.

One implication of these results is to call into question the suggestion made by Koerner (1989) that basal ice of the Greenland Ice Sheet is superimposed ice and that therefore intensive melting and perhaps the disappearance of this ice mass must have taken place during the last interglacial. The enrichment in ^{18}O of the basal ice, thought to be acquired during the formation of superimposed ice, may in fact be the result of isotope exchange with hydroxyl-bearing minerals.

When the dispersed and the stratified facies are present together at the same site, there is evidence that ice with dispersed debris lies above the stratified sequence and originates from a more distant source in the "interior" zone of the glacier. Drewry (1986) has recognized the widespread nature of this feature with the basal debris zone consisting of a lower stratified and an upper dispersed facies. Folding of basal ice in the marginal zone can, however, complicate the picture. It is tempting to visualize the origin of dispersed facies basal ice as the consequence of the onset of basal sliding. Downglacier, when more water is present at the base and when, closer to the margin, the ice-bedrock interface temperature can be below the melting point, the stratified facies can thus be formed beneath the dispersed facies by net accretion.

ıe Basal Zone of Alpine Glaciers

Processes operating in the basal zone of alpine glaciers are not fundamentally different from those acting under ice sheets and ice caps. However, the prevalent influence of water and the omnipresence of ice at its melting point temperature introduce special characteristics which can be considered separately. This is the main reason for this specific chapter.

5.1 Water Flow in the Basal Zone

In the Alpine glacial environment, the temperature of the ice is either at the melting point or very close to it. Water runoff at the surface originates from melting or as rain or dew. For a typical Alpine glacier, Röthlisberger and Lang (1987) consider that surface melt rates vary between 0.1 and 10 m/year depending on altitude while the frictional and geothermal melt rates are on the order of only 0.01 m/year.

In the accumulation zone, a snow or firn aquifer is usually present, from which water can drain by outflow at the firn line into channels on the glacier surface, by flow into crevasses and seepage through the glacier. In the ablation area, the water, having travelled some distance at the surface, usually disappears into the glacier.

Ice conduits exist throughout the glacier, the size of which reflects a balance between the melting of the ice due to the frictional heat produced by the flowing water, which tends to open them, and the plastic deformation of the ice resulting from the ice overburden pressure, which tends to close them. However, the gain or loss of energy which occurs when water adjusts to the pressure-dependent melting point is equally important. Drainage through large conduits originates where surface meltstreams disappear into moulins. Active moulins are usually located at the upstream end of crevassed zones, when the crevasses are merely cracks in the ice. From the observations that moulins are linked to cracks and crevasses and as water is denser than ice, it is conceivable that water-filled crevasses may propagate all the way to the glacier bed (Robin 1974; Weertman 1973). The propagation of a water-filled crack downwards is due to the increasing excess pressure (water pressure minus ice pressure) with depth. When the crack reaches the bed, sooner or later it will intersect a major subglacial channel and the water will drain.

Independently of ice conduits, the question of seepage and of the permeability of alpine glaciers must be considered. A drainage system exists through veins located at triple-grain boundaries although air bubbles tend to block them and ice deformation and recrystallization tend to close off many intergranular channels. An arborescent network of veins combining into tubes and then joining conduits has been envisaged by Shreve (1972). Raymond and Harrison (1975) calculated the distance travelled by such intergranular water corresponding to veins observed in ice cores from Blue Glacier, north-west United States. They concluded the maximum value to be 0.1 m/year, while Berner et al. (1978) deduced even lower maximum values in Griesgletscher in the Swiss Alps: between 0.02 and 0.04 m/year. They used the fact that, because some of the gases within ice are in contact with water, water seepage through ice changes the gas content and the gas composition. They measured how these quantities varied in surface samples along the central flow line of the glacier and, coupled with calculated travel times deduced from an ice flow model, the authors were thereby able to derive the water flux. These fluxes are very small compared with ablation rates, which are in the order of a few metres per year, concentrated in the summer. Thus, only a very small fraction of the surface meltwater penetrates the glacier by seepage through the vein system.

Water at the glacier bed may flow in different ways. Water may flow in channels incised into the substratum, called Nye-channels or N-channels. Water may flow in channels cut upwards into the ice. These are similar to conduits within the ice, except that they are at the ice-bed interface and are termed Röthlisberger channels or R-channels. Figure 5.1 gives the general configuration of N- and R-channels. Water drainage at the glacier bed occurs through an interconnected water system (Lliboutry 1987) which includes not only N- and R-channels but linked cavities (Walder 1986; Kamb 1987) and also a water film of variable thickness. The sheet flow of the basal water film primarily accommodates the local transport of melt waters associated with regelation sliding. However, water generated at the bed by geothermal heat and by the heat released by internal friction also flows in the film. Basal water discharge in a film requires an impermeable bed or a saturated subglacial aquifer. Water pressure in this layer is equal to the cryostatic pressure. On the upstream side of bed bumps, the water film thickness may be almost zero, while on the downstream side of these bumps, water-filled cavities can exist. The excess ice-pressure at the upstream side of obstacles plays a critical role in the development of the water film by closing water cavities and connecting exit channels, as can

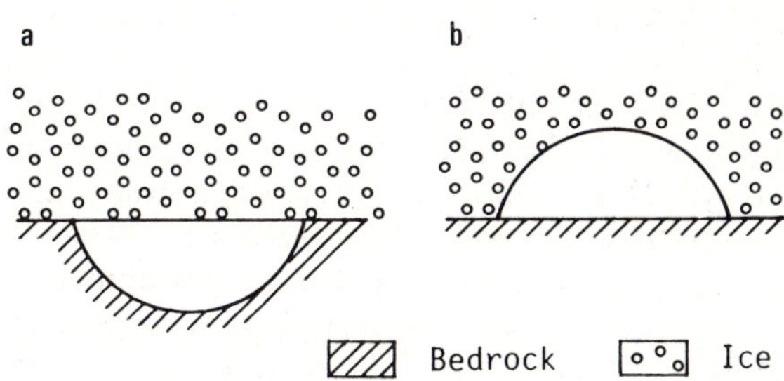

Fig. 5.1a, b. General configuration of N and R channels

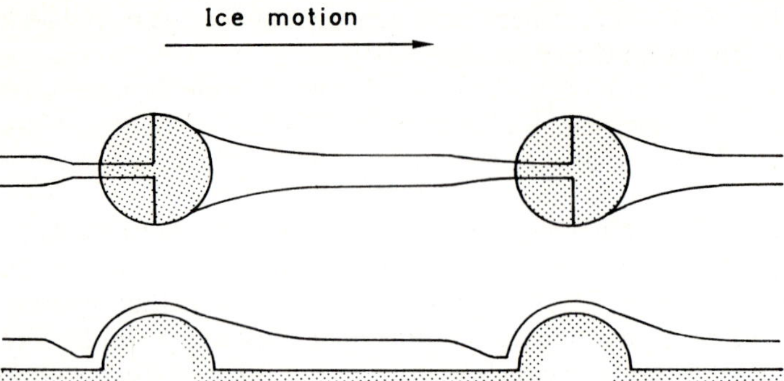

Fig. 5.2. Water cavities at the base of a glacier. *Above* Plan view of water cavities and connecting channels associated with a single obstacle-size glacier bed. *Below* Cross-section of water-flow path through water cavities and connecting channels. (After Figs. 2 and 3 in Weertman 1986)

be seen in Fig. 5.2 (Weertman 1986). With increasing water supply, instabilities might develop in the film and cause channelized flow (Walder 1982). Water films may become unstable when the thickness exceeds 4 mm; they will eventually turn into channels but the channels they turn into may themselves transform back into water films (Weertman and Birchfield 1983). This cyclic change need not take place simultaneously over the whole bed, but small areas that are in random phase with other local areas might cyclically switch from one form to the other. Water at the bed can be displaced by ice, unless the local melting rate equals the local creep-closure rate. Local melting from viscous dissipation of heat increases with channelization of flow. Creep closure increases with effective pressure. Thus the difference between the overburden pressure and the water pressure increases as flow becomes increasingly localized from films to linked cavities to channels (Alley 1989).

If attention is now focussed not on the water originating at the base of the glacier, but on surface meltwater, then this water remains concentrated in channels at the bed. Water arriving in substantial quantities at the sole of the glacier does not dissipate into a thin film but continues to flow in hydraulic channels at the bed. As a consequence large R-channels exist which are maintained at capacity, at least ephemerally, by the influx of surface meltwater.

Water-filled lee-side cavities play a significant role in glacier sliding (Lliboutry 1968) and the hydraulic action of water at high pressure in such cavities, associated with undulations or steps in the glacier bed, may explain the rapid displacement of Alpine Glaciers. Iken et al. (1979) have considered the effect of subglacial water pressure on the sliding velocity of Unteraargletscher in the Swiss Alps. Here, uplift of the glacier surface which accompanied enhanced horizontal displacement was considered to be a consequence of the formation of large water-filled cavities at the glacier bed. As indicated by Alley (1989), fast basal motion, whether by sliding or bed deformation, requires a water pressure close to overburden pressure and water pressure depends on water supply and drainage. Links exist between basal temperatures and surging

(Clarke 1976) and between hydrology and surging, but developments on these grounds are beyond the scope of this book.

5.2 Phase Changes at the Base of Alpine Glaciers

Englacial ice temperatures in an Alpine glacier are close to the pressure-melting point. Therefore, the ice-bed interface is generally at a temperature corresponding to the melting point and water is present at the base. The relative importance of basal sliding in Alpine glaciers is a direct consequence of this fact. Different sliding theories have been developed in the literature. These are reviewed by Paterson (1981) and the interested reader is referred to that book for further information.

With a bed at the pressure-melting point, temporal as well as spatial variations in basal pressure can result in phase changes. Observations by Kamb and LaChapelle (1964) beneath Blue Glacier in the north-western United States are most relevant in this respect. They illustrate the effect of a pressure change on the water film beneath only 26 m of ice. They report that, when overburden pressure is removed from ice in contact with bedrock by excavating the surrounding ice, the basal ice freezes fast to the rock. However, when excavated quickly without the initial release of overburden pressure, the ice comes free from the substrate and is not frozen to it. Robin (1976), commenting on this observation, shows that simple calculations indicate that a water film 10 μm in thickness requires about 0.08 cal of latent heat per cm^2 to freeze. This amount of heat could be supplied by conduction in about 15 s if sudden pressure removal results in the raising of the pressure-melting point by 0.1 °C.

As seen in Chapter 1, the pressure melting-regelation mechanism of sliding envisaged by Weertman is a consequence of the pressure variation from one side of a small protuberance on the glacier bed to the other. It gives rise to a regelation layer, the thickness of which, as suggested by Glen and quoted in Weertman (1964), is comparable to that of the controlling obstacle size. This reflects the fact that ice formed at the downglacier side of a smaller obstacle is likely to be melted completely when a larger (but still smaller than the controlling obstacle size) is encountered, while, on the other hand, if a bump larger than the controlling obstacle size is encountered, then the existing regelation layer will flow around it by plastic deformation in response to the local stress configuration.

Cold patches and water freezing can occur at the base of a glacier predominantly at the pressure-melting point as a consequence of the heat pump mechanism identified by Robin (1976). This can be viewed as another effect of the relatively rapid changes of pressure that occur as ice moves past an irregularity in the bedrock. Let us consider the effect of taking a sample of ice at the pressure-melting point in a perfectly insulated box and applying an additional hydrostatic pressure. The pressure-melting point will be lowered and thus, if any free water is present, the temperature of the ice-water interface will adjust to the new pressure-melting point. Heat will flow from the interior of the ice crystals to interfaces at grain boundaries and will be responsible for melting there. Since absorption of latent heat is a relatively efficient process, the temperature of the main mass of ice is thus lowered, as excess heat is used in melting the ice. In the upstream region of any bed protuberance, movement brings ice under increasing pressure, so that additional melting occurs. The water produced in veins or tubes along the triple junctions of crystals is likely to be squeezed out of the ice towards the subglacial water at lower pressure. Now, downglacier of the obstacle, the ambient pressure decreases in the parcel of ice considered. Unless there is some free water remaining in the ice that can release latent heat by freezing with this pressure decrease, then the ice remains at the temperature of the lowered pressure-melting point. Thus, as Robin suggests, this represents a simple heat pump effect that tends to cool the basal ice. The effect is associated with large bed protuberances as well as with small ones and is completely different from the pressure-melting-regelation mechanism. Cold patches are thus produced by the streaming of colder ice and the freezing of water at the interface can occur in these locations.

The effect described above and developed by Robin (1976) can take place in two cases: whenever ice moves from high to low pressure and whenever the pressure decreases because of a change in the stress distribution. This second case is envisaged by Röthlisberger and Iken (1981) in relation to water-pressure variations. During water-pressure induced changes in the stress distribution, patches of ice can freeze to the rock. The authors assume that water at high pressure has access to some cavities only, while others open passively with a concurrent drop in pressure, the freezing and adhesion of rocks being possible in the latter. Laboratory experiments have shown that small pressure fluctuations of only a few bars are sufficient for rock fragments to be frozen-on and lifted. Röthlisberger and Iken have developed a numerical model in order to understand the situation better. Of particular interest is the fact that, although the water pressure in the cavity may increase by only 0.7 bar, the ice pressure changes locally by some 13 bar, with the largest pressure drop located at the apex of the bed undulations.

Cold patches at the bed can result from thermal conduction through bedrock around areas of stress concentration. Figure 5.3 gives heat fluxes at the ice-rock boundary for a profile along a flow line; the ice motion is parallel to the dashed line, indicating the approximate upper limit of regelation ice. Thus, heat fluxes, corresponding to pressure melting on the upstream face and regelation downstream of the highest point, can be calculated at any point from the angle between the ice motion and rock surface. It can be seen that upstream of the rock crest two zones are present where the heat flux associated with pressure melting is greater than in the area between. This central section will be frozen due to conduction from the adjacent patches, although these patches are at the pressure-melting point.

Another possible way of providing the heat sink necessary for net basal freezing involves flow of air, caused by running water and the consequent aspirator effect, which would bring subzero air from the exterior of the glacier to basal cavities through crevasses, moulins and other englacial passages. The en-

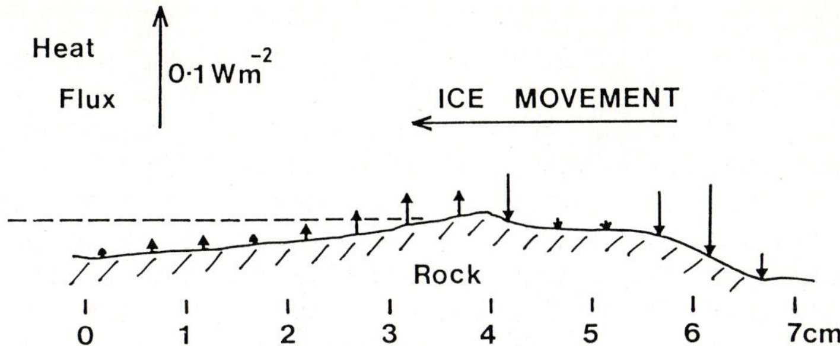

Fig. 5.3. Heat flux at ice-rock boundary for profile along a flow line. This assumes that ice motion throughout is *parallel to the dashed line* indicating the approximate upper limit of regelation ice. Numerical values of flux are *proportional to length of the arrow*, the scale being given by the sample arrow for heat flux 0.1 $W\,m^{-2}$. A heat flux towards the ice-rock interface indicates melting and vice versa. (Robin 1976, Fig. 3)

trainment of air by flowing water, a common air-flow mechanism, seems to be responsible for a substantial flow of air to the subglacial streams of alpine glaciers. Cavities in the marginal zones of alpine glaciers often are in direct contact with the cold air of the outside atmosphere, even if they are sealed by snow during the major part of the cold season. Refreezing of meltwater can thus take place in such an environment. This is also true for certain glaciers in the Alps, which exhibit cold ice at their base and subglacial permafrost along certain sections of their beds.

5.3 Incorporation of Debris into Basal Ice

Different mechanisms of debris incorporation can be envisaged to explain the basal ice characteristics encountered in alpine glaciers. The following example helps to clarify the problem.

Grubengletscher is located in central Wallis, Switzerland (Fig. 5.4) and flows from the crest of Fletschhorn (3996 m a.s.l.) in a north-west direction to an altitude of about 2800 m at its snout. Part of the glacier tongue and a nearby rock glacier are in a permafrost zone (Haeberli 1975) and temperature measurements made by Haeberli indicate that the ice at this point is clearly below the pressure-melting point. To control the level of an ice-dammed lake, a tunnel has been dug at the base of Grubengletscher, excavated partly in the frozen ground moraine and partly in the cold ice itself where it intersected a natural subglacial channel. Over a number of years, the tunnel showed little modification, but a small advance of the glacier has recently changed the stress field. The tunnel is closed at present but during the period in which it was accessible, a typical basal sequence was observed above the frozen ground moraine.

A closer examination of the frozen moraine reveals the importance of its silt content, the top 5 cm consisting in some places exclusively of fine silts. The occurrence of these fine particles can be related to water from the lake that cur-

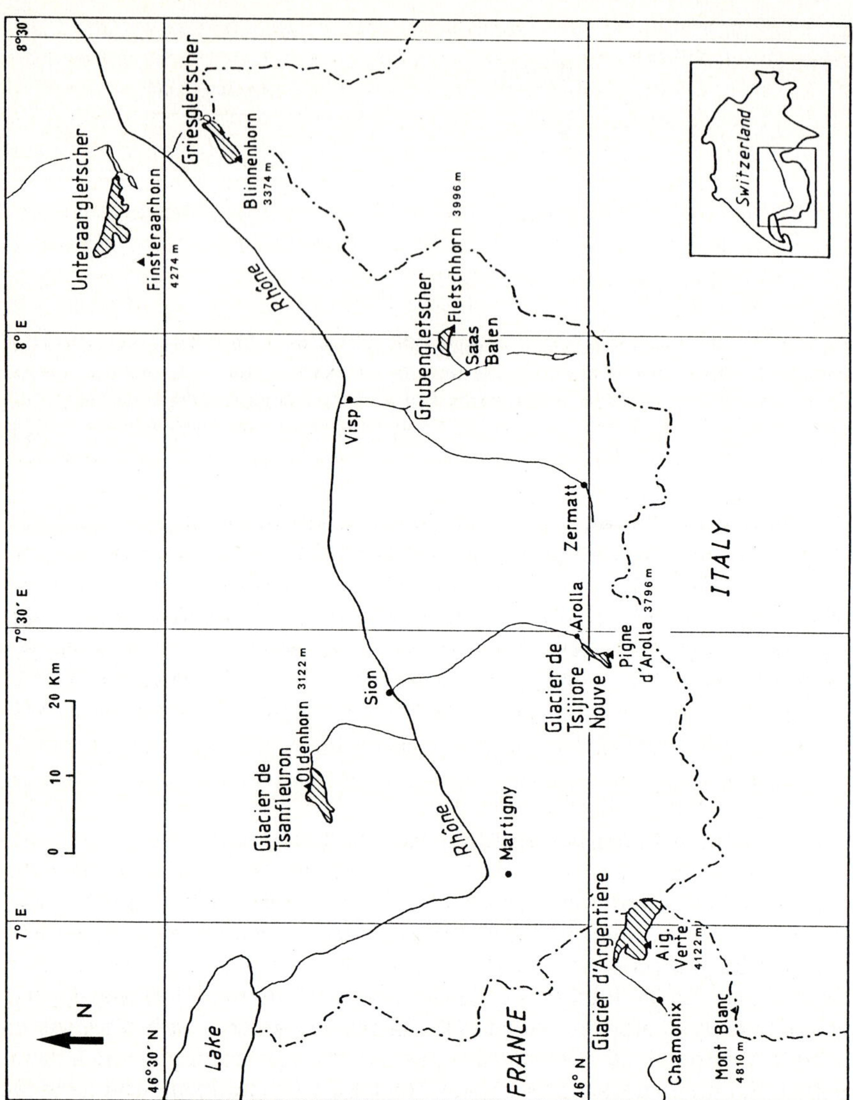

Fig. 5.4. Location map of the Alpine glaciers cited in this book

rently finds its way under the glacier and seeps through the upper part of the ground moraine. These fine silts are frozen and exhibit lenticular layers of clear ice. Tison et al. (1989) believe these features to be the product of ice segregation in thin layers in the upper part of the frozen ground moraine. Basal ice in the tunnel rises slowly towards the ice margin and it is intersected by two thrust surfaces about 5 m apart. One thrust surface contains a debris band that is connected to the ground moraine, while the other shows a debris band with lenticular ice layers that can be traced to the fine silts, again containing lenticular segregated ice layers, in a subglacial position. The latter continuity indicates that the material was frozen-on at the glacier base and transferred into the thrust by basal sliding. Due to ice deceleration and increased longitudinal compressive strain, such thrust planes may represent an effective means of transferring frozen soft sediments into an englacial position. As indicated by Drewry (1986), this process requires a longitudinal decrease of the rate of basal sliding. Besides, a certain amount of sliding and deformation of basal drift can occur at subfreezing temperatures as demonstrated by Echelmeyer and Zhongxiang (1987). The mechanism developed above also applies as subglacial temperature conditions change spatially in the course of time. In the case of Grubengletscher it should be remembered that the temperature at the ice-bed interface is only $-1\,^{\circ}C$. Under such thermal conditions, seeping of water from the ice-dammed lake through the upper part of the ground moraine, as was observed by the Swiss engineers in charge of the tunnel, may help to explain local sliding discontinuities and the formation of a thrust plane where the sliding ceases and the longitudinal compressive strain builds up. Fluctuating conditions at the glacier sole explain how the second thrust plane originated at a different location during another time period.

The example of the Grubengletscher tunnel illustrates an effective means of soft sediment removal under subglacial conditions. If the upper part of the sediment is frozen, it may be displaced with the glacier and eventually transferred into an englacial position along thrust planes resulting from ice deceleration at the base. If the bed is at the pressure-melting point, the permeability of the material considerably limits its erodibility. However, if the water pressure is high, erosion may occur by bed deformation rather than incorporation into the ice.

Incorporation of debris into the base of an alpine glacier usually depends on the formation of a basal ice layer due to phase changes at the glacier sole in response to various processes examined in the preceding section. This basal ice layer is characterized, as in polar ice caps and ice sheets, by a much larger debris content than glacier ice. The debris in the basal ice layer consists mainly of dirt or fine sand with occasional pebbles or boulders. The basal ice layer is generally stratified and is composed of a sequence of different types of layers of variable thicknesses. The following elementary layers constitute the basis of such sequences but are not always present together:

a) A bubble-poor ice layer about 0.2–3.0 cm thick, which is nearly particle free and is a single crystal layer thick.

b) A debris-rich layer of about the same thickness which is composed of fine mineral particles in an ice cement.
c) A very bubbly, but particle-free ice layer, the thickness of which is very variable from one sequence to another. This third type of layer is generally absent from the central bottom part of alpine glaciers and its formation seems to be related to its position along the marginal part of the glacier.

When boulders are present, the different layers forming the basal sequence are often interrupted at their contact rather than being deformed around them. A pocket of ice crystals is sometimes present in the lee of some boulders, indicating the filling by ice accretions of a small cavity.

Ice fabric analyses often fail to reflect the processes of formation of basal ice layers, since the pattern of the poles of c-axes on a Schmidt net is often not significantly different from the pattern obtained for glacier ice above. This is due to the fact that crystal orientation seems to adjust relatively quickly to the local stress field and no legacy of previous conditions is retained. Ice coatings formed on the floor of a subglacial cavity beneath Glacier de Tsanfleuron in the Swiss Alps have been studied by Tison and Lorrain (1987). The polished bedrock floor is coated with an ice layer 1 to 10 cm thick and, in one 8-cm-thick section, up to five debris layers, each 1 – 2 mm thick, have been observed. In places the ice coating is deformed and folded, giving rise to twisted sheets of ice. The ice coating is due to the partial freezing of water sheets entering the cavity during cold air penetration. A downward movement of the upglacier margin of the cavity drags the ice coatings along, detaches them from the bedrock and causes folding, while at the downglacier margin of the cavity, these folds are pinched between the glacier sole and the bedrock and become recumbent. This allows repetition of the debris layer sequence, which is further compressed under the glacier, while the prevailing stress conditions erase the initial fabric of the congelation ice. Figure 5.5 gives line drawings of crystal boundaries and fabric diagrams for the ice at various stages during this process; non-deformed ice coating, deforming ice coating and finally deformed ice coating incorporated at the sole. Glacier ice above the cavity is also given for comparison. The evolution involves the rapid disappearance of the fabric of the ice and an increase in crystal size.

5.4 Basal Ice Chemistry

The chemical composition of the ice constituting the basal ice layer can shed some light on the origin of the water refreezing at the ice-bed interface. The case of Glacier de Tsijiore Nouve, located on gneissic bedrock, is worth considering in this respect. On the eastern margin of the glacier, just below an ice fall (Fig. 5.6), marginal crevasses and a tunnel allowed Souchez and Tison (1981) to study the basal ice layer in detail. Figure 5.7 gives the alkali versus the alkaline earth content of different types of ice from this glacier, each point on the diagram representing one sample. Samples of snow, some samples of

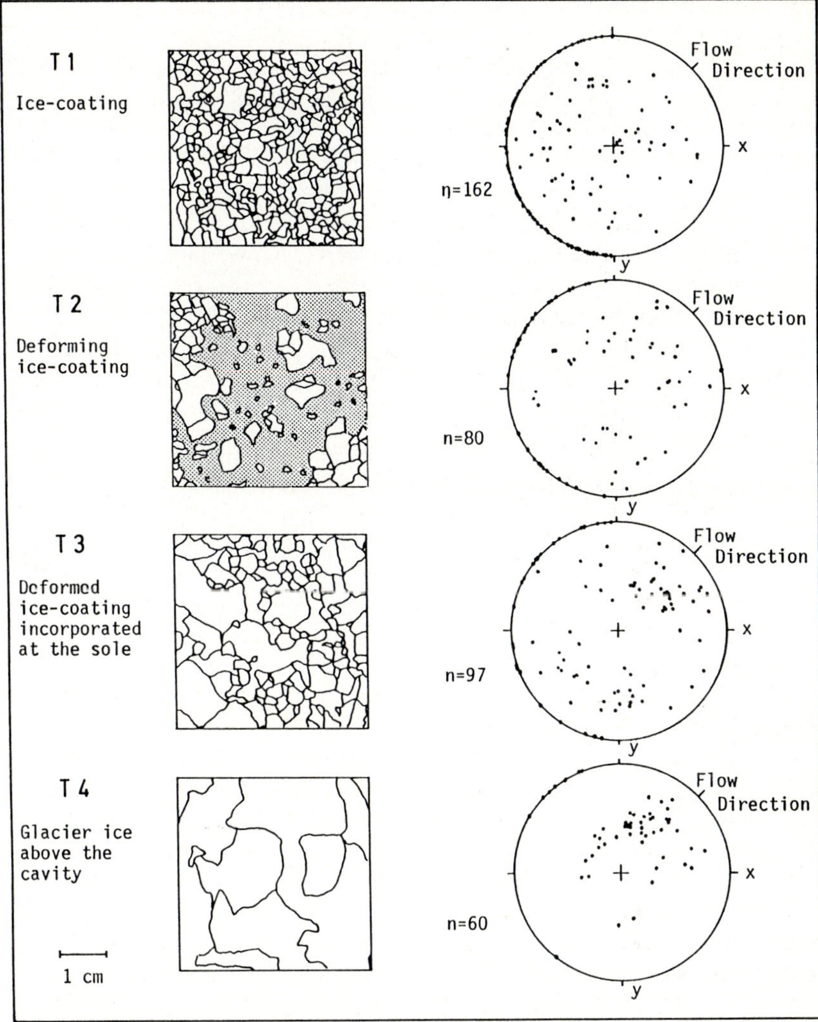

Fig. 5.5. Line drawings of crystal boundaries and fabric diagrams of the different types of ice samples in a subglacial cavity of Glacier de Tsanfleuron, Swiss Alps. The thin sections were cut parallel to the bedrock. *n* is the number of crystals measured. Because of software requirements, all the equatorial optic axes are plotted on the left-hand side of the diagrams. The *dotted area* in T2 represents the microcrystalline ice matrix. (Tison and Lorrain 1987, Fig. 5)

glacier ice and most samples of basal ice at the margin of the glacier are located in the lower left part of the diagram. They represent, from a chemical point of view, a single population. Two trends can be observed. If, during ice formation, prolonged contact with rock particles occurs, the (Na+K)/(Ca+Mg) ratio is greatly reduced because of a preferential contribution of alkaline earths by the particles. This is indicated by the very low (Na+K)/(Ca+Mg) ratio in

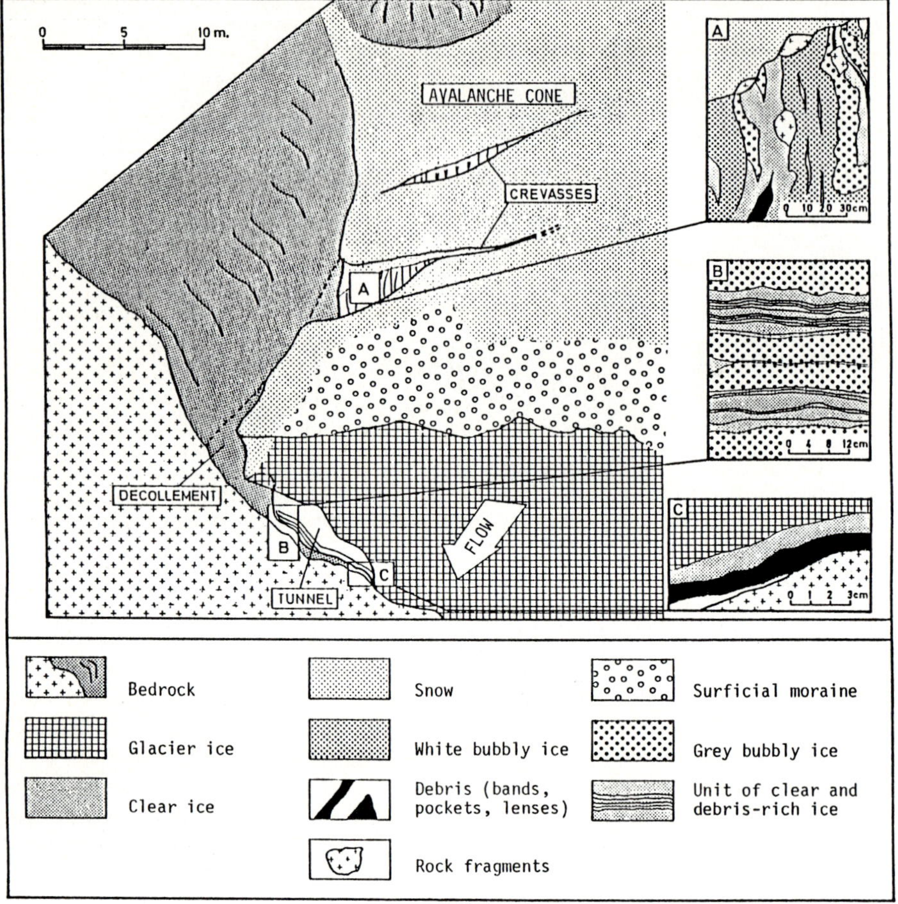

Fig. 5.6. Detailed sketch and vertical cross-sections near the eastern margin of Glacier de Tsijiore Nouve. (Souchez and Tison 1981, Fig. 1)

subglacial flowing water (0.05 – 0.09) and in supraglacial streams (0.11 – 0.14) and by the following experiment. Fine particles (< 50 μm) of morainic deposits from the Glacier de Tsijiore Nouve were left in contact with melted ice from the same glacier at 0 °C and the water was sampled over the period from 1 – 320 min. It was observed that the (Na + K)/(Ca + Mg) ratio decreased with increasing contact time (Fig. 5.8), a value of 0.18 being recorded after 1 min of contact and a value of 0.05 after 320 min. Some glacier ice and pond ice samples fall in this category, as indicated by the lower arrow in Fig. 5.7. On the other hand, Souchez et al. (1978) have shown that a selective flushing-out of ions occurs as a consequence of water squeezing and that ice formed from refreezing of this water has a much higher (Na + K)/(Ca + Mg) ratio. This trend is represented by the upper arrow in Fig. 5.7. The basal bubble-poor ice in-

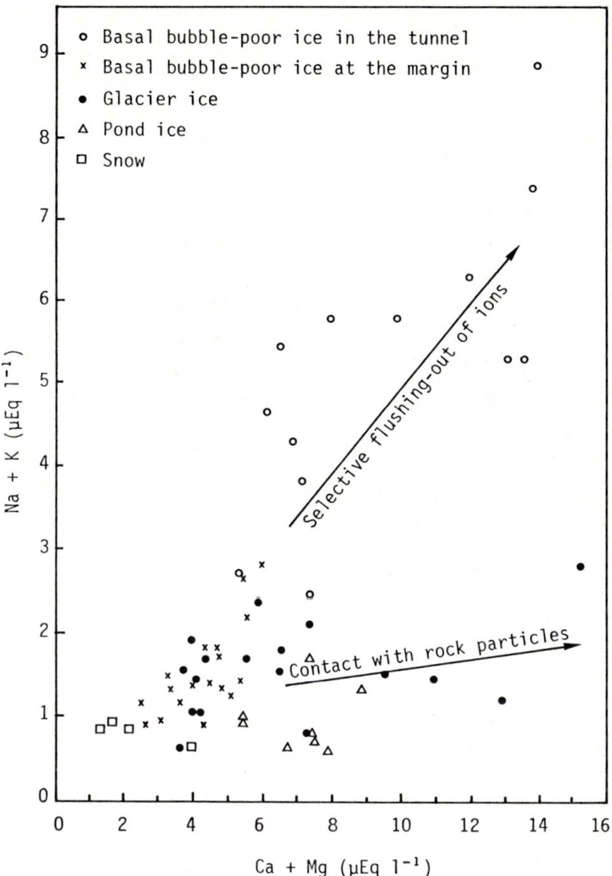

Fig. 5.7. Relationship between alkali and alkaline earth content of the types of ice studied from Glacier de Tsijiore Nouve. (Souchez and Tison 1981, Fig. 2)

vestigated in the subglacial tunnel falls into this category, showing (Na+K)/(Ca+Mg) ratios varying from 0.34 to 0.85 with a mean value of 0.58. A fractional melting experiment carried out on glacier ice samples from Glacier de Tsijiore Nouve shows results along the same lines as mentioned in Section 2.6. Pressure melting within the basal ice mass may be responsible for the formation of excess water in high-pressure zones and this water can be squeezed out of the ice along the pressure gradient. This is connected with the heat pump effect mentioned in Section 5.2.

The formation of the basal ice layer upglacier from the tunnel investigated at the Glacier de Tsijiore Nouve is related to refreezing of water squeezed out from the ice. Indeed, meltwater circulating at the base has a low (Na+K)/(Ca+Mg) ratio and this will decrease further during freezing, since divalent ions are preferentially incorporated into the growing ice, as shown by Malo and

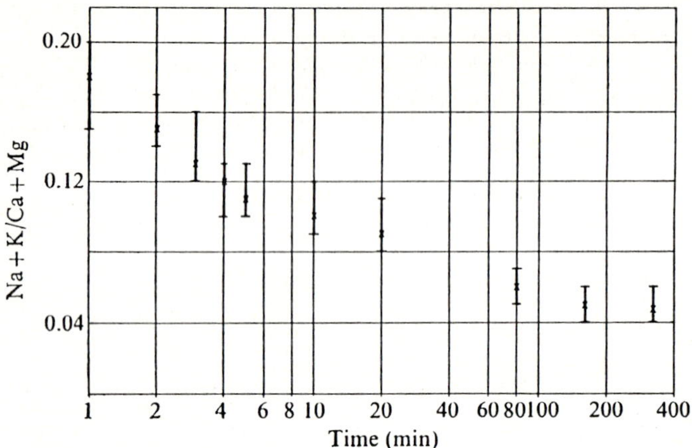

Fig. 5.8. Ratio of alkalis to alkaline earths as a function of time in the experiment. Time in this figure is the time elapsed between the contact of the first drop with the particles to the complete filtering. (Souchez et al. 1978, Fig. 2)

Baker (1968). The chemical composition of basal, bubble-poor ice in the tunnel implies squeezing of water from the ice and only limited contact with rock particles. Since this kind of ice only exists as lenses above a debris band visible in the basal ice layer of the tunnel, it is assumed that, at first, squeezed water invades the pores of the subglacial permeable sediment and refreezes. Then this frozen layer acts as a barrier, allowing the formation of lenses of debris-free, bubble-poor ice above. Ion exchange between water and particles is thus greatly reduced. The situation is discussed by Souchez and Tison (1981).

In certain circumstances, particle layers may act as semi-permeable membranes, as demonstrated by Souchez and Lorrain (1975) for a frozen mud layer plastered on the ceiling of an enclosed natural subglacial cavity beneath Glacier d'Argentière (French Alps). A pressure gradient exists across this mud layer: interstitial water in the glacier at its upper surface is under a hydrostatic pressure corresponding to about 90 m of ice and, its lower surface, 2 to 3 cm below, is under the atmospheric pressure of the cavity. This interstitial water is thus forced through the mud layer and then refreezes, due to the pressure dependence of freezing point, in the form of ice accretions hanging from the ceiling. A comparison between the chemical composition of samples of the ice accretions and water held at their surface (S1) and samples of water in contact with the upper surface of the mud layer (S2) is made in Table 5.1. The following significant characteristics can be pointed out:

a) A lower content of all four cations in the water sampled at (S1) than at (S2), showing a desalinating effect as water is forced from (S2) to (S1).
b) A higher (Na+K)/(Ca+Mg) ratio for (S1) than for (S2), indicating a monovalent-divalent cation species separation.

Table 5.1. Concentrations of Na, K, Ca and Mg in the samples collected at the base of the Glacier d'Argentière, French Alps. (Souchez and Lorrain 1975, Table 1)

Ionic species or ratio	Water in contact with the mud layer μEq/l (17 samples)			Ice accretions and water held at their surface μEq/l (9 samples)		
	Mean	Min.	Max.	Mean	Min.	Max.
Na	71.6	28.3	139.2	55.8	34.8	80.0
K	176.4	61.4	296.6	58.7	35.3	81.8
Ca	460.0	249.5	774.4	38.9	6.5	64.4
Mg	50.0	17.3	95.4	17.6	9.1	32.9
Na+K+Ca+Mg	758.0	371.5	1263.7	171.0	89.7	242.2
(Na+K)/(Ca+Mg)	0.50	0.21	0.90	2.35	1.50	4.59
Na/K	0.41	0.23	0.61	0.96	0.81	1.12
Ca/Mg	10.02	6.49	15.83	2.37	0.49	5.51

c) A higher Na/K and a lower Ca/Mg ratio for (S1) than for (S2), indicating sorting effects within the alkalis and within the alkaline earths as well as between the alkalis and the alkaline earths.

These characteristics indicate that the frozen mud layer plastered at the base of the glacier may be considered as an ion-exchange membrane causing electrolyte filtration and relative retardation of cations as described by Helfferich (1962). The mud layer is frozen at a temperature close to the melting point, with the result that large pores are filled with ice and that unfrozen water exists partly at the interface between mineral grains and ice and partly in crevices and capillaries where the radius of curvature is sufficiently small (Anderson 1967). In such a situation, Hoekstra and Chamberlain (1964) have shown that the ions must migrate along thin films of unfrozen water which are continuous through the frozen mud. The membrane characteristics indicated above are explained by the peculiar electrochemical properties of these films, whereby a "Donnan exclusion" exists and a selective electrolyte filtration effect occurs. Indeed, Kemper et al. (1970) and Berry (1969) have indicated that with a low hydraulic pressure gradient, such as that existing across the frozen mud layer at the base of the Glacier d'Argentière, one can observe the following relative retardation sequence: Na < K < Mg < Ca. This sequence implies a higher (Na+K)/(Ca+Mg) ratio, a higher Na/K ratio and lower Ca/Mg ratio for the effluent solution, which is matched exactly at the base of Glacier d'Argentière. A chemical sorting effect is thus taking place at the base of this Alpine glacier.

Study of the lead 210 activity of the basal ice layer (B.I.L.) from the southern margin of Glacier de Tsanfleuron in the Swiss Alps has enabled the identification of a marginal accretion process. The southern margin of this glacier exists as a relatively gentle slope and, from place to place, the glacier loses contact with the bedrock. Some natural subglacial cavities are thus accessible and give opportunities for close observation of the B.I.L. It consists of a wedge of ice the thickness of which diminishes gradually towards the interior, being 1

Table 5.2. ^{210}Pb Activity of ice samples from the right side of Glacier de Tsanfleuron, Swiss Alps. (Souchez and Lorrain 1978, Table 3)

	dpm/kg
Glacier ice	0.454 ± 0.03
	0.501 ± 0.03
	0.700 ± 0.05
Basal ice layer	0.196 ± 0.03
near the entrance of the tunnel	0.210 ± 0.03
	< 0.03
Basal ice layer	4.262 ± 0.20
near the end of the tunnel	1.009 ± 0.05
	2.264 ± 0.10

or 2 m thick at the glacier margin and disappearing completely after about 8 to 10 m. The thick bubbly layers of this B. I. L. were examined by Souchez and Lorrain (1978).

The ^{210}Pb activity, in dpm/kg of ice, of the B. I. L. is compared in Table 5.2 with the activity of immediately overlying glacier ice. Percolation of melt water through small crevices in the ice can modify the ^{210}Pb activity by radionuclide migration in the water and considering such alpine glacier ice as a closed system may be an approximation. However, the internal consistency of the results in Table 5.2 signifies that this approximation is reasonable for the southern side of the Glacier de Tsanfleuron tongue. The filtration role played by the particles in the ice, which tend to absorb migrating radionuclides (Prantl et al. 1973), must also be considered. This effect can be discounted in the present case, since particles were carefully removed prior to sample melting, and since dirt extracted from a known volume of ice showed only negligible ^{210}Pb activity.

Table 5.2 also shows that values of activity for glacier ice are well grouped around the mean, whereas for the wedge of basal ice below, two groups can be distinguished: low values of activity are encountered near the margin, probably because a greater influence of melt water, and high values – higher than for the glacier ice above – are obtained from samples near the end of the subglacial tunnel. Melt water may have a highly variable ^{210}Pb activity depending both on its origin (old glacier ice, recent snowbank) and on the relative mobility of this radionuclide during partial melting.

The fact that basal ice at the inner extremity of the wedge has a substantially higher ^{210}Pb activity than glacier ice immediately above substantiates an origin for that ice by basal accretion. Blowing snow deposited in a marginal crevasse under glacier ice is a normal feature of this margin of Glacier de Tsanfleuron and by compression induced by the overlying glacier ice, blowing snow, rock fragments, and regelating melt water may form several basal layers in the course of time. Such a B. I .L. is evidently much more recent than the glacier ice above and this age difference explains the contrast in ^{210}Pb activity. Some values encountered in the B. I. L. reached the ^{210}Pb activity of a fresh snowfall (mean value: 4 dpm/kg), indicating that the accretion process described was operating at the time of study.

Subglacial meltwater on the upglacier side of a bed obstacle has a solute content equal to that of the basal ice plus any contribution from the weathering of the bedrock. As indicated by Hallet (1976), in the case where the solute content of basal ice is uniform, the water that freezes on the downglacier side of a bed protuberance has an average composition equal to that of the water on the upglacier side divided by the effective distribution coefficient. As this coefficient is less than unity, the solute concentration in water on the downglacier side of the obstacle is generally higher than that on the upglacier side. As the freezing temperature of water is relatively sensitive to the presence of solutes, this difference in the subglacial solute concentration will influence the regelation sliding process. Specifically, a higher solute content tends to lower the freezing point, reducing the temperature gradient across bedrock obstacles and thereby inhibiting the heat flow that drives the regelation process. An idea of the importance of this phenomenon can be gained from the calculations of Hallet which indicate that the temperature drop resulting from an excess pressure of one bar on the upglacier side of an obstacle would be nullified by an excess of 1.2×10^{-3} mol of dissolved $CaCO_3$ on the downglacier side. Thus, accumulation of solutes along lee surfaces tends to impede regelation slip and thereby glacier sliding, particularly when the bed surface is characterized by roughness elements whose mean heights do not exceed about 1 m. For obstacles of greater size, enhanced creep plays the major role in allowing ice to pass round them.

5.5 Subglacial Precipitates and Basal Ice

Subglacial calcite precipitates are usually observed as patchy coatings, up to a few centimetres thick, developed on outcrops that have been recently exposed by retreating alpine glaciers. These carbonate deposits, often present on polished and striated bedrock pavements, are characteristically fluted and furrowed parallel to the former ice flow direction. They are finely laminated, each individual layer being up to 100 μm thick, and they are preferentially developed on the lee side of small rock protuberances. They have been described from the Alps (Bauer 1961; Lemmens et al. 1982) and from the Canadian Rockies (Ford et al. 1970; Hallet 1976) and have also been observed on the floors of subglacial cavities (Souchez and Lemmens 1985).

Calcite precipitation is generally the consequence of a loss of carbon dioxide from solution, a common process in caves where the partial pressure of carbon dioxide in influent-saturated solutions of calcium carbonate is greater than in the cave atmosphere. However, the loss of carbon dioxide is not the only means by which calcite precipitation occurs. A solution of calcium carbonate can become saturated or oversaturated if partial freezing is taking place since, during the growth of ice, ions are mostly rejected into the remaining liquid phase. The subglacial environment is a place where such freezing may occur as a consequence of large pressure fluctuations in association with the pressure dependence of the freezing point of water. It follows that the subglacial environment is a likely place for calcite deposition.

Hallet (1976) has made a significant contribution to the problem of the formation of these subglacial calcite precipitates. These deposits appear to form in intimate association with the pressure-melting and regelation process, which is an important component of glacier sliding. In this process, basal ice melts because of relatively high pressures along the upglacier surfaces of bedrock protuberances. The water so produced corrodes the limestone rock and flows as a thin subglacial film along the pressure gradient to neighbouring lee side surfaces, where it refreezes because of the decrease in pressure. During this freezing, solutes present in water tend to be preferentially rejected by the growing ice and therefore tend to accumulate in the liquid phase. By repeating the melting and refreezing process on further bedrock protuberances, this mechanism of concentrating solutes may be sufficiently effective to increase the concentration to the eutectic point, defined by simultaneous ice formation at the base of the glacier and calcium carbonate precipitation at the interface. The eutectic point for calcium carbonate is close to 0 °C so that the process may occur at the base of a glacier which is at the pressure-melting point at its base. To test this mode of formation, thin layers of bubble-poor basal ice have been sampled by the present authors for their chemical composition at the Glacier de Tsanfleuron, Switzerland, a present-day site of formation of these carbonate deposits (Hallet et al. 1978).

The Ca concentration in basal ice from the Glacier de Tsanfleuron is estimated by calculating the eutectic Ca concentration. The eutectic Ca concentration in solution, expressed in Eq/l can be adequately represented by the following equation:

$$Ca = 0.0282\ (pCO_2)^{0.365}\ ,$$

where the carbon dioxide partial pressure, pCO_2, is expressed in bar. The method of estimating the effective pCO_2 consists of:

a) demonstrating that, gas bubbles being present in a limited amount in the basal ice examined, the subglacial waters are in equilibrium with the gas phase,
b) evaluating the most plausible values for the composition of gases in these bubbles, and finally
c) arriving at the effective pCO_2 by multiplying the estimated pressure in subglacial waters by the relative abundance of CO_2 in bubbles present in these waters. If the gaseous species of bubbles in regelating waters are present in the same relative abundances as in regelation ice, then freezing subglacial water ought to contain about 1% CO_2 by volume. The mean pressure in the subglacial water film is 5.3 bar (based on ice thickness) and so the estimated pCO_2 is 0.053 bar, which corresponds to a eutectic Ca concentration in solution of 9700 μEq/l.

After estimating the eutectic Ca concentration in subglacial waters, an effective distribution coefficient is needed to predict the Ca content of regelation ice. Experiments in which $CaCO_3$ solutions have been frozen at rates and with freezing front morphologies believed to be comparable with those expected in regelation sliding, show that ice tends to have a solute content 50 to 100 times smaller than the melt (Hallet 1976). Using a distribution coefficient of 0.01 and a range of total pressure between 0.75 bar when the ice loses contact with the bed (0.75 instead of 1 bar because of the altitude of the site) and 5.3 bar, calculated eutectic Ca concentrations in ice range from 47 to 97 μEq/l. This expected range of Ca values overlaps almost exactly with the measured Ca concentration in basal ice, which varies from 45 to 106 μEq/l. Thus, the Ca content of ice from the base of the Glacier de Tsanfleuron can be successfully predicted from considerations of the phase relations in regelating subglacial solutions in approximate equilibrium with subglacial calcite deposits.

Souchez and Lorrain (1978) have made calculations to establish the minimum Ca concentration in basal ice for the case of regelation in equilibrium with calcite. For a region of about 2500 m a.s.l. they obtained a value of 13 μEq Ca/l. Basal ice from numerous sites in the Alps, such as Glacier de Tsijiore Nouve for example (Table 5.3), does not reach such a concentration level and it therefore appears that their chemical composition is not controlled by the solute redistribution that accompanies regelation. As seen in Section 5.4, in the case of Glacier de Tsijiore Nouve, exchangeable ion reactions dominate the basal ice chemistry.

An isotopic study of basal ice, subglacial water and the subglacial carbonate deposits of the Glacier de Tsanfleuron sheds more light on the processes involved. In this area, the calcite samples collected near the glacier margin have a mean $\delta^{18}O_{PDB}$ value of −6.77‰ with extreme values of −5.58 and −7.72. The limestone bedrock on which the calcite is precipitated has a mean $\delta^{18}O_{PDB}$ value of −4.02‰ with extreme values of −3.29 and −4.41. This difference indicates that calcite found in the deposit is not simply derived from the bedrock and that the carbonate has probably undergone an equilibration

Table 5.3. Cationic composition of different types of ice from Glacier de Tsijiore Nouve, Swiss Alps. (Souchez and Lorrain 1978, Table 1)

	Glacier ice (18 samples) $\mu Eq\,l^{-1}$			Basal bubbly ice (11 samples) $\mu Eq\,l^{-1}$			Basal bubble-poor ice (13 samples) $\mu Eq\,l^{-1}$		
	Mean	Min.	Max.	Mean	Min.	Max.	Mean	Min.	Max.
Na	0.8	0.3	2.0	1.4	0.5	2.3	2.9	1.3	5.2
K	0.9	0.3	1.4	1.1	0.1	2.4	2.4	0.8	3.7
Ca	4.2	1.9	9.6	4.1	2.2	7.3	6.4	3.4	10.7
Mg	2.6	1.4	7.1	2.7	2.1	4.3	3.1	1.9	4.4
Na+K+Ca+Mg	8.5	4.3	18.0	9.2	5.1	15.1	14.8	8.0	22.8
(Na+K)/(Ca+Mg)	0.28	0.09	0.49	0.34	0.15	0.51	0.58	0.34	0.85

of its oxygen isotopes with water of different isotopic composition. On the other hand, the similar $\delta^{13}C_{PDB}$ values of the two materials – around –2‰ in both cases – implies that no organic carbon was involved. Now, precipitation must have occurred in isotopic equilibrium with water since it is due to a concentration increase by freezing. It is thus accompanied by a preferential incorporation of ^{18}O in the calcite corresponding to a fractionation factor of 1.0347 for calcite and water at 0 °C (Clayton et al. 1968). Using this factor and the calculated $\delta^{18}O_{SMOW}$ of the subglacial calcite, one obtains $\delta^{18}O_{SMOW}$ = –10.92‰ on the average for subglacial water in equilibrium with it. Extreme values obtained by the same method are –9.74 and –11.87‰ (Lemmens et al. 1982; Souchez and Lemmens 1985). Now, water in equilibrium with calcite is the residual water that has been enriched in $CaCO_3$ by partial freezing. It has been shown in Chapter 2 that, during freezing, water in equilibrium with the growing ice becomes progressively impoverished in heavy isotopes. Therefore, the residual water in equilibrium with the calcite considered here must have a lower $\delta^{18}O_{SMOW}$ value than the initial subglacial water at the onset of freezing. Thus, initial water or melted ice –there is no fractionation on melting of glacier ice – giving rise to the precipitate by freezing must have a $\delta^{18}O_{SMOW}$ value higher than –10.92‰. The only isotopic values from this environment which are not too negative to be at the origin of the initial water are those of basal ice. Thus, the initial water from which calcium carbonate is precipitated by partial freezing can only originate from melting of the basal ice layer. The meltwater thus produced will refreeze in subsequent lee side locations, dissolving limestone during each cycle, and will finally give rise to a calcite coating on the bedrock surface by precipitation at the eutectic point. Meltwater from glacier ice cannot be at the origin of the process and the use of this deposit as a palaeoclimatic indicator to reconstruct the isotopic composition of former glacier ice as envisaged by Hanshaw and Hallet (1978) is therefore questionable.

Subglacial precipitates other than calcite also occur in recently deglaciated areas. Hallet (1975) reports silica-rich precipitates; iron-oxide and aragonite deposits (Aharon 1988) have also been observed.

5.6 Isotopes in the Basal Zone of Alpine Glaciers

An oxygen isotope investigation of the basal zone of Matanuska Glacier in southern Alaska has been conducted by Lawson and Kulla (1978). Figure 5.9 gives the $\delta^{18}O$ values of ice samples of the different facies recognized by the authors along a vertical profile. The "diffused" facies is essentially a facies of glacier ice containing angular debris entrained superficially in the accumulation area. As in other glaciers, the basal zone consists of an upper "dispersed" facies and a lower "stratified" facies. The ice of the stratified facies shows an enrichment in oxygen 18 compared with ice from the facies above and freezing-on is believed to be the main process by which this facies originates. It must, however, be remembered (see Sect. 4.3) that a study of a single isotopic ratio

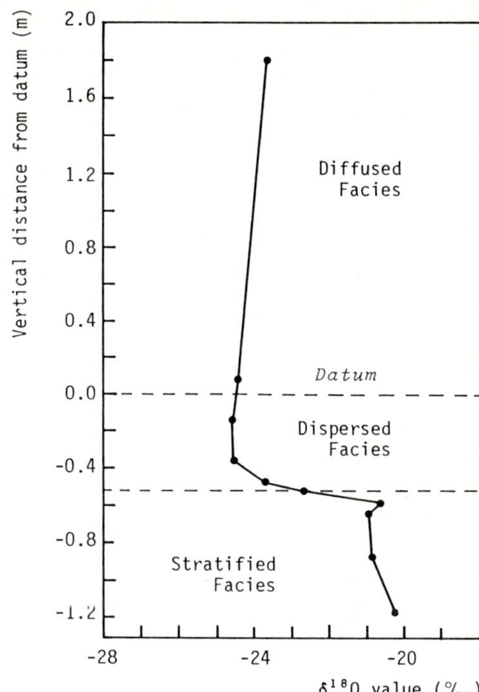

Fig. 5.9. Comparison of $\delta^{18}O$ values of ice samples of the diffused, dispersed and stratified facies from a transect of the frontal zone of Matanuska Glacier (Alaska). (Lawson 1979, Fig. 25)

cannot give a full explanation, since the amount of fractionation which takes place during freezing is not known. Even in the simplest situation, for example the freezing of water in a closed system in isotopic equilibrium allowing the application of the Rayleigh model, ice layers can be either isotopically enriched at the beginning of the process or isotopically impoverished at the end, compared with the initial reservoir. Under natural conditions the problem is further complicated by non-equilibrium processes, the trapping of liquid water, and input and output of water from the reservoir. Thus it is not possible to show whether a freezing process has taken place in a reservoir from the measurement of only one isotope. The δD-$\delta^{18}O$ slope is required to characterize the freezing process, as indicated in Chapter 2.

Using the equation for open system freezing [Eq. (6) in Sect. 2.4], the calculated freezing slope for an Alpine glacier having $\delta_i D = -90‰$ and $\delta_i{}^{18}O = -12.5‰$ is 6.39. The δ values chosen in this example are typical for the Alps, so that the freezing slope here is closer to the precipitation slope than in polar regions. Figure 5.10 shows a δD-$\delta^{18}O$ diagram for Glacier de Tsanfleuron ice samples. A freezing slope is well displayed for the stratified basal ice, a situation which appears to be the result of melting without isotopic change and refreezing with isotopic fractionation. The differentiation occurs because the isotopic fractionation coefficients are such that the $(\alpha-1)/(\beta-1)$ ratio is different from 8, α and β being the equilibrium fractionation coefficients for deuterium and oxygen 18 respectively. The freezing slope decreases

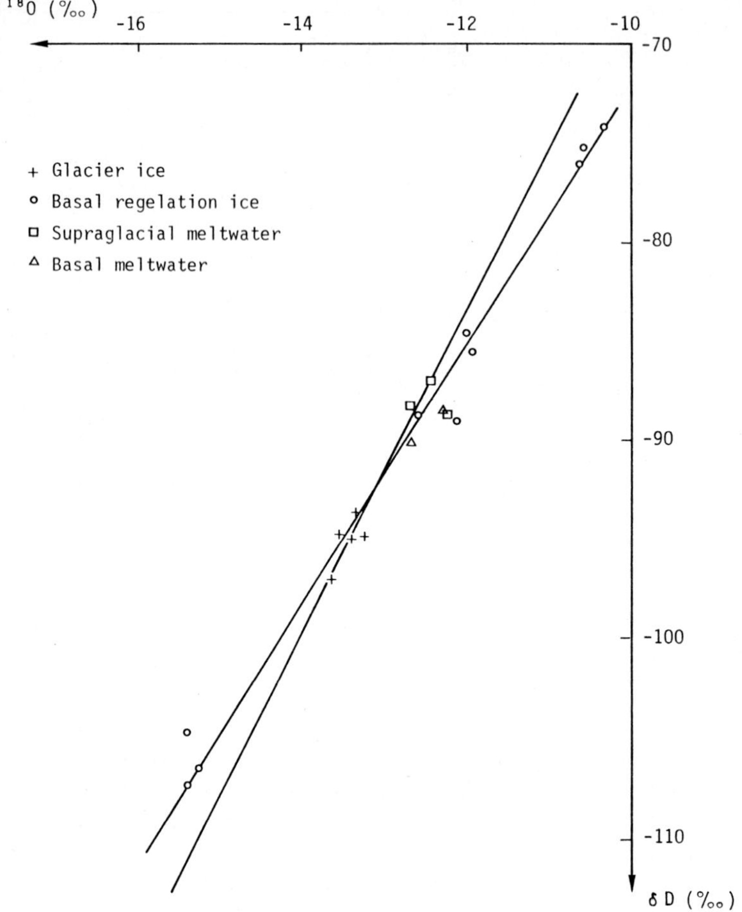

Fig. 5.10. δD-δ^{18}O diagram for Glacier de Tsanfleuron ice samples. (Lemmens et al. 1982, Fig. 5)

as the initial isotopic composition in δD and δ^{18}O becomes more negative; this is the reason for the slope difference between alpine and polar glaciers.

The freezing slopes in open and closed systems are practically the same if the isotopic composition of the input to the open system is not significantly different from that of the initial reservoir. As outlined in Chapter 2, the theory has been developed to include the evolution of the freezing slope during mixing with a more negative water as input. The results of the computer simulation can explain the situation encountered at the base of Grubengletscher in the Swiss Alps.

The site and the subglacial tunnel are described in Section 5.3. Glacier ice, basal ice and water samples are given in the δD-δ^{18}O diagram in Fig. 5.11. All the samples presented in this figure show a linear relationship with a slope of 8.23 and a correlation coefficient of 0.993. However, a more careful study of

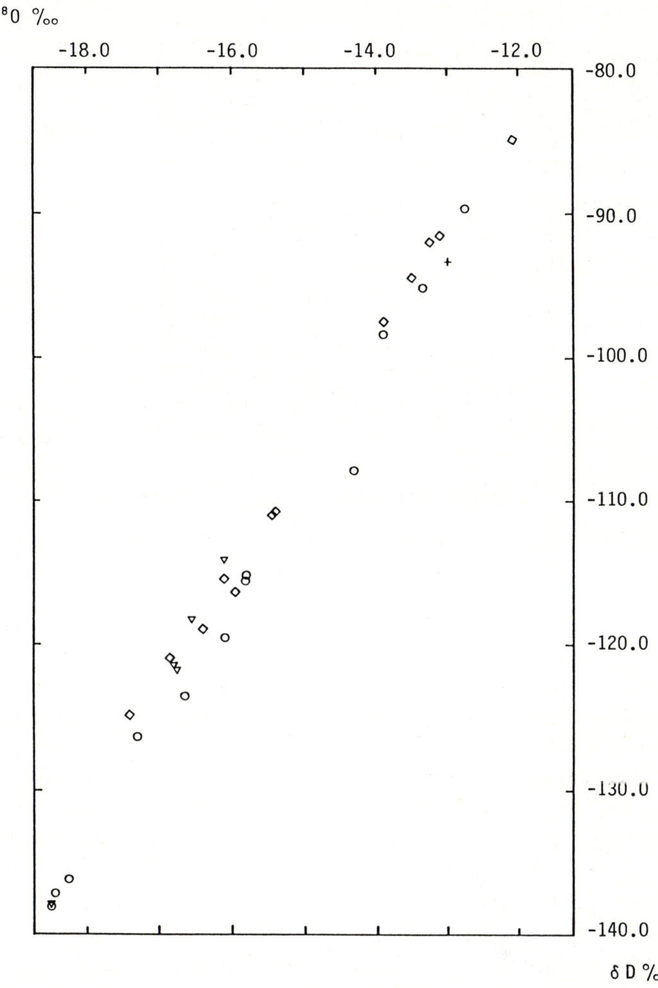

Fig. 5.11. δD-δ^{18}O diagram for Grubengletscher ice and water samples. *Open circles* for basal ice samples, *open squares* for glacier ice samples, and *open triangles* for water samples from the ice-dammed lake. (Souchez and De Groote 1985, Fig. 1)

ice samples leads to the recognition of two populations with different deuterium excess values. The mean value of deuterium excess for glacier ice is 13.0, while it is 10.5 for basal ice. Distribution of the numbers of samples versus deuterium excess classes for glacier ice and basal ice is indicated in Fig. 5.12. Separate slope computations for glacier ice and for basal ice samples give similar values of 7.96 and 8.19 respectively with a correlation coefficient greater than 0.995. No distinction can be made between these two groups in terms of slope and only slight differences in deuterium-excess values exist. The results can be explained if a source of more negative water than the initial water is

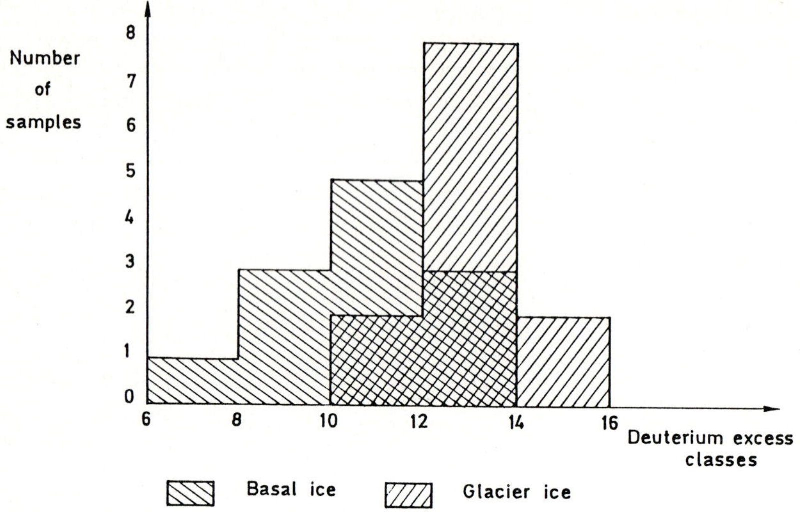

Fig. 5.12. Distribution of deuterium excess values in glacier ice and basal ice samples from Grubengletscher. (Souchez and De Groote 1985, Fig. 2)

present. Souchez and De Groote (1985) consider that the ice-dammed lake located in the proximity of the tunnel where basal ice samples were obtained, could be the source of this input to the subglacial freezing process. The lowest isotopic values of the ice-dammed lake indicated on Fig. 5.11 were obtained during late autumn when the upper part of the lake was already frozen while the highest δ values indicated for the lake were obtained during the summer after a heavy rain storm characterized by surprisingly high δ values. Water from the lake currently finds its way under the glacier and seeps through the upper part of the ground moraine. Thus it is quite probable that in winter this lake water contributes to the subglacial water that is allowed to freeze in an open system at the glacier sole. Let us consider that initial water has $\delta D = -114‰$ and $\delta^{18}O = -16‰$ (represented by a point on Fig. 5.11) and that the input is water from the ice-dammed lake at the lowest isotopic values measured. The calculated pin point (represented by a cross on Fig. 5.11) will have $\delta D = -93.2‰$ and $\delta^{18}O = -13‰$.

The computer simulation indicates that the open system model with an input poorer in heavy isotopes than the initial reservoir will lead to a progressive change of slope around a pin-point that can be calculated from the equations

$$\delta D = \delta_i D + 20.8 \quad \text{and}$$
$$\delta^{18}O = \delta_i^{18}O + 3 \ ,$$

where δ_i values are δ values of the initial water, $\alpha = 1.0208$ and $\beta = 1.003$. Basal ice samples can be aligned on a precipitation slope with a slightly lower deuterium excess if an input of water impoverished in heavy isotopes is mixed with initial water during the course of freezing.

The position of the calculated pinpoint on Fig. 5.11 close to the least negative basal ice fits with such a mixing model. Similarly, the difference in deuterium excess between glacier ice and basal ice predicted by the model is 3.2 as compared with 2.5, from the Grubengletscher samples, again in good general agreement.

The Grubengletscher case study reflects a situation where water impoverished in heavy isotope is mixed as input with initial water in the course of freezing. In such a case basal ice samples may be aligned on a steeper slope than the freezing slope and, eventually, no distinction can be made in terms of slope from glacier-ice samples. Only the deuterium excess classes for the two populations will be different, as shown in Fig. 5.12 for Grubengletscher.

6 The Contact Zone Between Glacier and Ocean

6.1 Ice Shelves and Tidewater Glaciers

Ice shelves are floating ice sheets attached either to land or to a grounded ice sheet. They represent seaward extensions of terrestrial glaciers in the majority of cases. A substantial part of the Antarctic coast line is fringed by ice shelves, while they are also present on a reduced scale in the Northern Hemisphere, for example the Ward Hunt Ice Shelf along the northern coast of Ellesmere Island in Arctic Canada. Ice shelves are nourished by snow accumulation on their upper surface, by flow of ice from adjoining ice masses and also by bottom freezing at the interface with sea water. Wastage is caused by ablation at the surface, which is in most cases not very significant, by bottom melting at the lower interface in contact with sea water and by calving at the frontal zone which produces icebergs. Where an ice shelf grounds locally it slows down and tends to thicken, resulting in an ice rise with its own flow pattern determined by grounded-ice dynamics.

The calving process is common to ice shelves and tidewater glaciers, both of which have their outer limit afloat. At the front of a floating glacier, which for simplicity is assumed vertical, the normal stress is equal to the water pressure and varies linearly from zero at water level to $\varrho_w g h'$ at the bottom, where ϱ_w is the water density, g is the acceleration of gravity and h′ the distance from the base of the glacier to the water surface. At equilibrium, this normal stress must equal the cryostatic pressure. If the glacier has a thickness h, the cryostatic pressure at the base $\varrho_i g h$ (where ϱ_i is the density of the ice) must be equal to $\varrho_w g h'$. Thus $h' = \varrho_i/\varrho_w h$. In Fig. 6.1 from Reeh (1968) it can be seen that, comparing this stress distribution with that necessary for keeping the ice in equilibrium, the actual stresses are insufficient to maintain equilibrium. The deviation is a tensile force N and the ice must expand in the direction

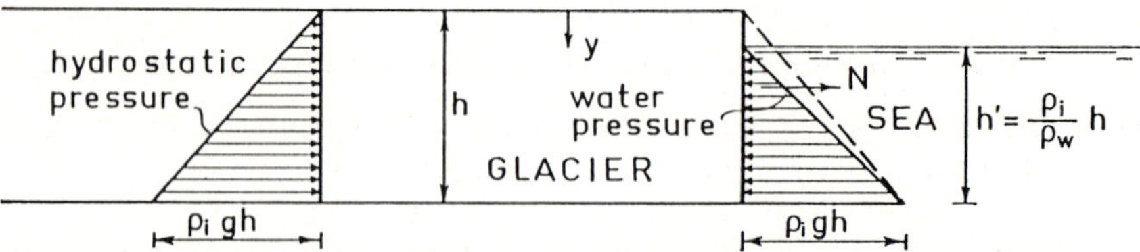

Fig. 6.1. Longitudinal section of a floating glacier (see text). (Reeh 1968, Fig. 1)

of this tensile force, perpendicular to the front. Since the deviation between the actual pressure and the hydrostatic pressure increases at the front of the glacier from the bottom to the top, the floating glacier or ice shelf is subject to bending and shear stresses develop in the ice body. Reeh (1968) indicates that the tensile and shear stresses reach their maximum in a cross-section situated at a distance from the ice front approximately equal to the thickness of the glacier. Since this combination of tensile and shear stresses is likely to induce ice fracture, icebergs about the length of the ice thickness are likely to be produced. Although this process is considered to be important, it is not the only mechanism responsible for iceberg calving. Many small icebergs are produced by pieces of ice falling from the upper part of the ice front.

Large stresses are produced by buoyancy effects arising from tidal variations, and these are concentrated at the grounding line separating the grounded part of the glacier from the floating part. However, the discrepancy between observed iceberg size and the horizontal distance between the grounding line and the ice front seems to indicate that these effects are not important in the calving process. The action of storm waves at the ice front does not generate stresses which can lead to calving, but the waves do induce pressure fluctuations at the lower surface of the floating glacier. These pressure fluctuations decrease rapidly with distance below the sea surface and are no longer of importance at a depth equal to the wave length. Thus the action of storm waves can only be considered as an effective process of calving in thin floating glaciers, since wave lengths for ocean waves of more than 100 m are exceptional. Tsunami-like storm waves are a contributing factor, while creep failure due to lateral spreading is another failure mechanism reflective of interaction between the ice shelf and environmental factors. Lateral stresses are generated in the ice as a result of the spreading of ice shelves under their own weight. Robin (1979) regards this process as probably the most common form of calving in the Antarctic, but it is difficult to distinguish this effect from that resulting from the imbalance of hydrostatic forces along the vertical face of the front as pointed out above (Reeh 1968). A useful review of iceberg calving and deterioration in Antarctica is given in Kristensen (1983).

Ice shelf equilibrium reflects a balance between a number of factors. For instance, an ice shelf may thicken in response to any of the following causes (Thomas 1979):

a) An increase in drainage from the ice sheet.
b) An increase in snowfall on the adjoining ice sheet causing growth.
c) A reduction in creep rates by alteration of ice-shelf dynamics.
d) An increase in snowfall on the ice shelf.
e) A decrease in bottom-melting rates or increase in bottom-freezing rates.

Thickening of the ice shelf will itself change the position of the grounding line and this will affect flow in the ice sheet; such complex interactions are being developed and studied intensively at the present time.

Since there is no friction at the interface between the ice shelf and the ocean, the driving force of the weight of ice above sea level is balanced by the restraining force due to shear at the sides. The floating tongue of a tidewater glacier shows a decrease in ice thickness down the inlet. This results from the drop in the horizontal pressure due to the shear stress on the side walls and the consequent creep of the floating ice. As indicated by Crary (1966), this decrease is inversely proportional to the width of the valley, the ice thickness and volume of outflow having only secondary effects. For similar reasons, the longitudinal thickness gradient is inversely proportional to the width of an ice shelf. Thus, ice shelves are characterized by a flat upper surface and by a thickening of the ice as one goes from the frontal cliff towards the grounding line.

Since ice shelves rest on a frictionless bed, i.e. ocean water, velocity and strain-rate measurements made at the surface approximate closely those at depth. Let us consider a band of ice shelf bounded laterally by flow lines. As indicated by Thomas (1979), the volume of ice entering and leaving the band can be calculated from measurements of ice velocity and thickness, while the volume of ice added by snow accumulation is obtained from measurements of net accumulation rate. If the ice shelf is in the ablation zone, the volume of ice lost is obtained from measurements of net ablation rate. Imbalance between calculated volumes gives the ice-shelf thinning or thickening rate. This can be due to bottom melting or bottom freezing which is an important question in the study of the contact zone between ocean and glacier. This question will be now considered.

6.2 Melting and Freezing at the Base of Ice Shelves

That phase changes occur at the sole of an ice shelf has been indicated by the fact that the temperature of sea water near the ice shelf, at a depth corresponding to its lower boundary, can be lower than the melting temperature of the ice constituting the ice shelf. In contrast, sea water temperatures around −1.8 °C have been measured, i.e. a temperature about 0.2 °C higher than the freezing point of sea water, taking into account the salinity.

If the temperature of sea water is above its freezing point, basal ice will be melted until the freezing point is reached for water at a certain level of salinity in contact with the sole of the ice shelf (Doake 1976). The salt-water-ice system diagram of Fig. 6.2 explains the situation. The curve L-E gives the reduction in freezing point temperature with increasing salt concentration from L, the freezing point of pure water to E, the eutectic temperature. Similarly the curve S-E gives the equilibrium line for salt precipitation from the solution. Removal of water by freezing increases the salt concentration so that solutions submitted to partial freezing follow the curve L-E called the "liquidus". Let us now suppose ice and sea water at the same temperature, but lower than the melting temperature of the ice constituting the ice shelf, i.e. less than 0 °C, and above the sea water-freezing temperature. For equilibrium to be obtained on the liquidus, ice must melt to reduce the salinity of sea water in contact with it and the temperature of this sea water must be lowered in order to provide the latent heat necessary to melt the ice. The final state of the system is given by a point

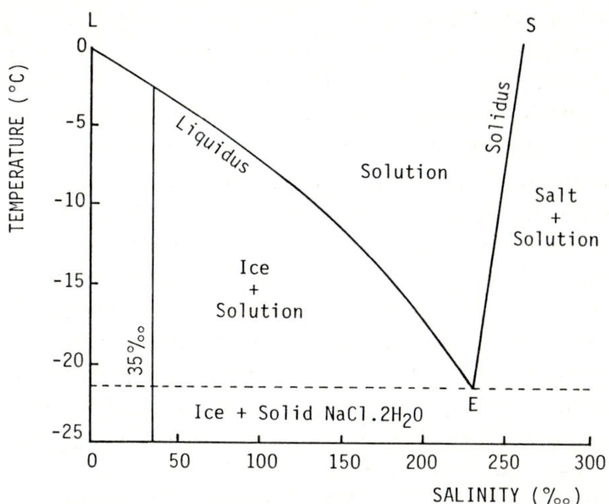

Fig. 6.2. Ice-water-salt system phase diagram. *E* = Eutectic point. (After Fig. 13.4 in Drewry 1986)

on the liquidus. Since the salinity of sea water is about 35‰ and the eutectic point corresponds to a salinity of 230‰ the S-E curve or solidus does not have to be considered here.

Doake (1976) also gives the condition for freezing-on where sea water must already be at its freezing point, i.e. represented by a point on the liquidus, and the ice must be at a lower temperature. Freezing takes place at the ice-water interface and the latent heat is removed by conduction through the ice at a rate dependent on the temperature gradient. At the same time, the sea water at the contact will become more saline as salts are rejected from the growing ice phase. Conduction of latent heat raises the temperature of the ice until it reaches that of sea water when freezing will cease.

Pressure can affect the thermal regime at the base of an ice shelf. If water flows in the upglacier direction at the base of an ice shelf, it will descend, since the ice thickness increases and the ice base is at a lower level. The increased pressure will result in lowering of the freezing point so the descending water can only reach equilibrium by melting the overlying ice and lowering its temperature. Robin (1979) discusses the reverse process: if a water current ascends from a deeper level, the freezing point of the water can be reached and freezing can thereby occur at the base of an ice shelf. A 5-mm layer of ice can be produced if a cubic metre of sea water is elevated by 500 m. Considering an irregular ice-shelf base with thicker ice corresponding to ice streams entering the shelf, a current flowing along the bottom will be able to melt ice at the base of the thicker ice stream extensions and to accrete ice in the intervening zones of thinner ice.

The temperature field within ice shelves can be investigated during borehole drilling and the form of the temperature-depth curve is partly determined by the basal heat flux. First, the englacial temperature field can be used as an indicator of basal melting or freezing as developed by Zotikov (1986). Let us consider a steady-state ice shelf with a thickness assumed to be constant in the course of time. If there is no vertical movement of ice, the temperature distribution with depth is a straight line between the equilibrium temperature with sea water at the base and the mean annual surface temperature at the upper boundary. A vertical downward movement of ice results from accumulation at the upper boundary and melting at the base of the ice shelf. In such a circumstance, cooling of the ice occurs and, at each level, the temperature will be lower than that occurring with the same boundary conditions in the absence of vertical movement. The temperature-depth curve is now concave-upwards, with higher thermal gradients in the basal part of the ice shelf. On the other hand, a vertical upward movement of the ice results from freezing at the base and ablation at the surface. Warming of the ice is caused by such an effect and, at each level, the temperature will be higher than that occurring with the same boundary conditions in the absence of vertical movement. The temperature-depth curve in this case is concave-downwards, with lower thermal gradients at the base of the ice shelf.

Temperature distributions recorded in the ice shelf near Maudheim and in the Ross Ice Shelf near Little America V station are both examples of concave-upward curves indicative of basal melting (Fig. 6.3). In contrast, the temperature distribution in Koettlitz Ice Tongue, a floating glacier in McMurdo Sound, exhibits an upward convexity below the zone affected by seasonal temperature changes thereby indicating bottom freezing (Fig. 6.4). On the Amery Ice Shelf, a third type of temperature distribution is found. It is an S-shaped distribution corresponding to the combined effect of surface accumulation and freezing at the base (Fig. 6.5).

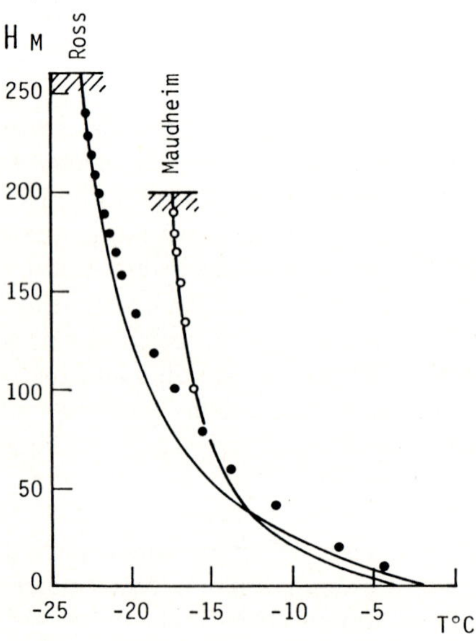

Fig. 6.3. Temperature distribution in two Antarctic ice shelves. (Zotikov 1986, Fig. 8.2)

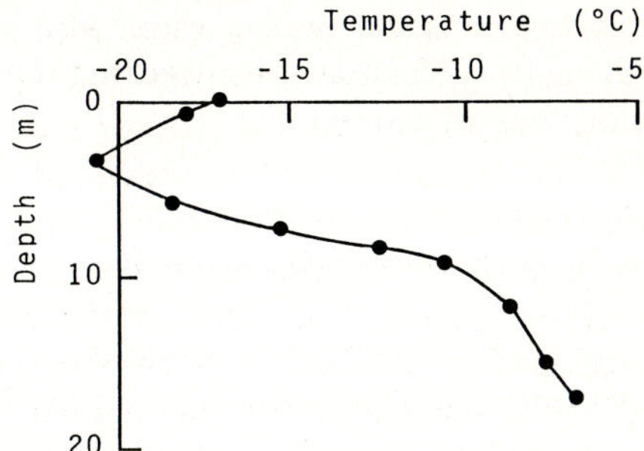

Fig. 6.4. Temperature distribution in the Koettlitz Ice Tongue, Antarctica. (After Fig. 9.7 in Zotikov 1986)

Temperature distribution modelling, using known rates of ablation or accumulation at the surface and various melting or freezing rates at the base, results in the generation of a series of curves. It is possible to match a measured temperature distribution with one of these curves in order to estimate the melting or freezing rate at the base of the ice shelf concerned. An average melting rate for the coastal strip of Antarctic ice shelves is about 30 cm/year, with values up to 60 cm/year in some places (Zotikov 1986), while freezing rates are much lower, in the order of 3 to 5 cm/year under present-day conditions.

However, in some circumstances, fresh water rather than sea water is present at the base of the ice shelf. This water comes from beneath the continental glaciers and is discharged into the sea at the grounding line. Due to its lower

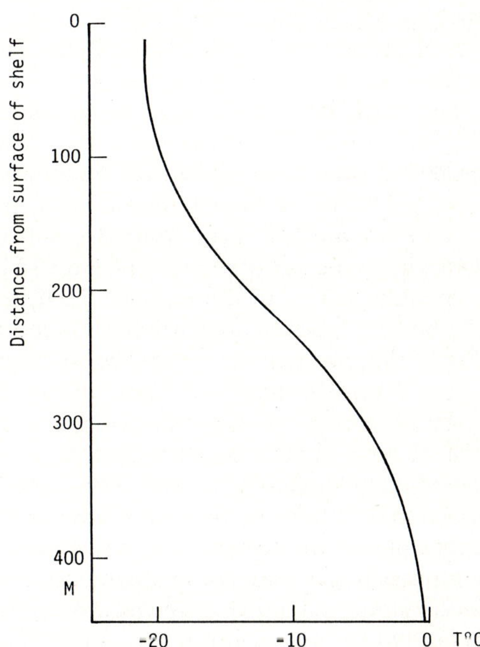

Fig. 6.5. Temperature distribution in the Amery Ice Shelf. (Zotikov 1986, Fig. 9.5)

density, it floats on sea water and is thus in contact with the sole of the ice shelf. In such situations, freezing rates can increase from a few cm/year to several tens of cm/year.

6.3 Frazil and Congelation Ice

Two types of sea ice, congelation ice and frazil ice, are worth considering here in connection with the freezing-on of sea water at the base of an ice shelf.

Congelation ice is formed by the direct freezing of sea water to the bottom of an existing ice sheet due to the downward migration of a freezing front through sea water. Congelation sea ice can be identified by the presence of a characteristic cellular structure consisting of more or less evenly spaced ice platelets or "cells", separated by small angle grain boundaries. Salt is concentrated as brine inclusions along the boundaries and adjacent layers of brine pockets are separated from each other by the "brine layer spacing", measured parallel to the c-axis. Laboratory experiments have shown that ice formed from sea water under conditions of unidirectional freezing has a brine layer spacing that is dependent on the freezing rate. As indicated by Weeks and Ackley (1986), when the growth rate is lowered, the average brine layer spacing increases. When a congelation sea ice layer is well developed, elongated crystals are aligned in the vertical with c-axes close to horizontal and it is for this reason that this zone is called the columnar zone. During the advance of the freezing front, salt becomes entrapped in the ice phase, largely in the form of brine inclusions located along the boundaries between the ice plates. This entrapment occurs because sea ice forms along a non-planar interface, in turn a product of pronounced supercooling in the water layer ahead of the advancing interface. If supercooling occurs, an unstable situation may develop where any segment of the interface that advances ahead of the rest reaches a region of greater supercooling where it experiences a greater driving force and a still faster growth rate. Steep-walled cell boundary grooves associated with the entrapment of brine are thereby produced at the interface.

An undamaged core from the bottom of the Ross Ice Shelf has been brought to the surface close to Camp J9 in the central inner part of the shelf. The undersurface of the Ross Ice Shelf in this area was found to be experiencing bottom freezing. Zotikov (1986) reports that the undamaged undersurface shows tiny protrusions composed of the rounded ends of vertical ice crystals about 5 mm in diameter. These protrusions are not randomly arranged but form horizontal, parallel rows (Fig. 6.6). In fact, the entire lower 6-m layer of the ice core is made up of such crystals and resembles congelation sea ice as observed in an annual sea ice cover. There is therefore little doubt that the progression of a freezing front through sea water at the base of the ice shelf is responsible for the structure. The spacing between the vertical platelets is about ten times larger than the analogous characteristics in annual sea ice, implying a very slow growth rate at the base of the ice shelf. Weeks and Gow (1978) consider that the alignment of crystals at the base of an ice cover results from the

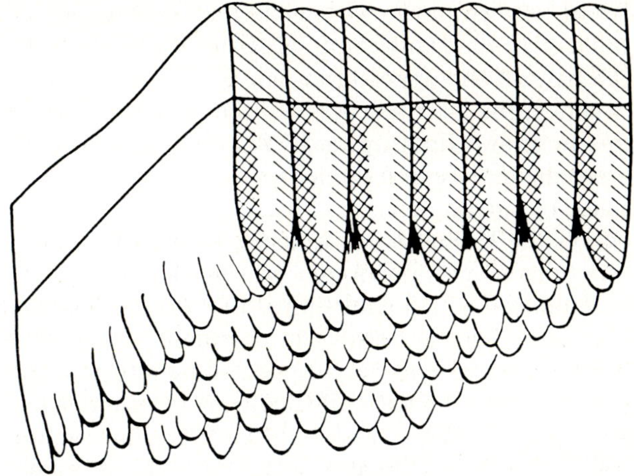

Fig. 6.6. Alignment of vertical crystals discovered at the bottom of the Ross Ice Shelf. (Zotikov 1986, Fig. 9.12)

presence of a current in the water layer beneath. The mechanism operates through the control exerted by the current on the composition of the liquid at the dendrite tips and thus on the interface temperature. Consequently, the existence of crystals oriented in horizontal rows at the base of the Ross Ice Shelf at J9 can be considered as an indication of the presence of a prevailing sea current beneath the ice shelf.

Frazil ice consists of ice crystals nucleated within the ocean and agglomerated at the sea surface as a sponge-like body. When agglomerated, the resulting ice layer is composed of fine-grained, equigranular crystals with randomly oriented c-axes. Turbulence, induced by wind and waves, can be a factor leading to frazil ice formation; but this effect is limited in leads and polynias and near ice edges, due to the extreme damping when a significant percentage of the sea surface is covered by ice. Beneath the ice shelf, water which is at a depressed freezing point due to pressure at depth can be brought adiabatically to the surface of the ocean. This can occur near the front of large ice shelves where frazil ice may form as a result of an increase in freezing point temperature. However, only a small quantity of ice results from such a mechanism, since the depression of the freezing point by pressure is only a few tenths of 1 °C. The contact between two water masses of significantly different salinities but both at their freezing point, represents an important mode of frazil ice generation. As indicated by Weeks and Ackley (1986), freezing occurs due to double diffusion where the transfer of heat takes place at a faster rate than that of salt. Terwilliger and Dizio (1970) report that the diffusion coefficient of NaCl in water is in the order of 10^{-5} cm^2/s and since the thermal diffusivities of ice and water are in the order of 10^{-2} and 10^{-3} cm^2/s respectively, heat transfer occurs much more readily in this system than mass transfer. If a layer of water at lower salinity lies on top of more saline water, both at their freezing point, the transfer of heat from the warmer upper layer to the colder layer beneath is more rapid than the transfer of salt from the lower to the upper layer. As a result, the base of the upper layer can be the site of a phase change. Ice nucleation occurs at the interface between the fresher and the more saline water; ice crystals form and float upwards. Another situation where double diffusion plays a prominent role in frazil ice generation is that of thermohaline convection initiated by surface freezing. Freezing of the upper part of the oceanic mixed layer entails the formation of a dense, salty, cold water layer beneath the ice front due to salt rejection by the growing ice. A descending cold brine plume can be produced which enters into contact with less cold and less saline surrounding waters. The descending brine plume gains heat but loses salt at a much lower rate, thereby cooling adjacent waters to temperatures below their freezing point. Thus, ice crystals nucleate and rise under the influence of buoyancy. As developed by Weeks and Ackley (1986), convection is continuously strengthened by such ice formation, with the effect of producing a convection cascade. This process explains the generality of frazil formation under thick ice and is thus likely to occur under an ice shelf where bottom freezing occurs.

Frazil ice often contains fine-grained sediment or algae. The reason is either because these particles act as nuclei for frazil crystal nucleation or because they are captured during the upward movement of frazil crystals. The two processes can easily be distinguished: the foreign particle is the center of an ice crystal in the first case and is at the boundary between crystals in the second.

6.4 Isotope and Impurity Distribution

The isotopic and chemical composition of ice from ice shelves is a good indicator of flow characteristics, of glacial supply from the adjacent continent and of the occurrence of freezing at their base. Concerning the latter process, an oxygen isotope profile of basal ice is of great interest. Sea water has an isotopic composition close to SMOW. If freezing occurs, the maximum enrichment possible in oxygen 18 of the ice due to the value of the equilibrium fractionation coefficient is 3‰. If sea water has a $\delta^{18}O = 0$, the maximum $\delta^{18}O$ value of ice resulting from the freezing of that water is +3‰. By contrast, snow accumulated at the surface of the ice shelf and ice flowing from the adjacent continent have very negative $\delta^{18}O$ values since they represent the end of the atmospheric water cycle depleted in heavy isotopes. The contrast between very negative $\delta^{18}O$ values for ice having an atmospheric origin and $\delta^{18}O$ values close to SMOW for ice resulting from sea water freezing therefore represents a significant tool in the discrimination of ice origins in an ice shelf.

The example of a core drilled in the middle of the Amery Ice Shelf serves to illustrate this. A borehole 315 m deep has been drilled in the shelf at the point marked G_1 on Fig. 6.7a. The $\delta^{18}O$ profile at G_1 suggests, as indicated by Morgan (1972), that the shelf consists of ice from three distinct sources. The uppermost layers, from 0 to 70 m, have $\delta^{18}O$ values from −19 to −23‰ (Fig. 6.7b) typical of snows deposited near the coast at low elevation and it is

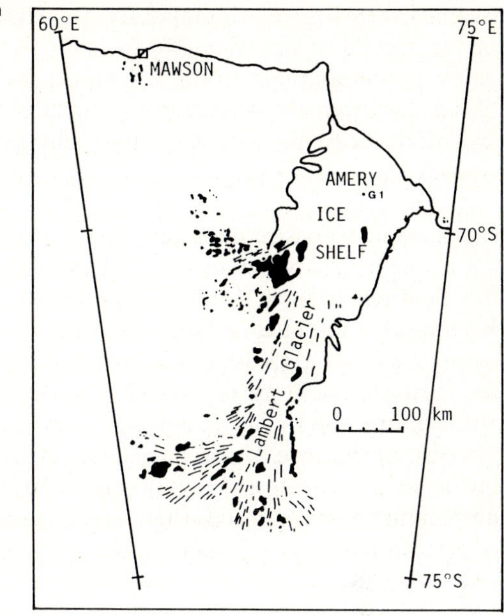

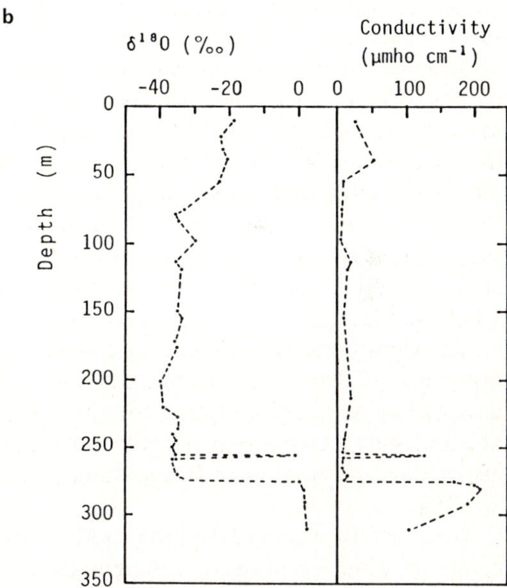

Fig. 6.7. The Amery Ice Shelf. **a** Position of the borehole (*G1*) and lines of principal ice flow in the Lambert Glacier. **b** Oxygen isotope (*left*) and electrical conductivity (*right*) profiles to a depth of 315 m at location *G1* in the ice shelf. (Morgan 1972, Figs. 1 and 2)

assumed that this ice originated from precipitation onto the ice shelf itself. The second section from 70 to 270 m has $\delta^{18}O$ values between −35 and −40‰, which are comparable to those of the Lambert glacier drainage basin further inland. Ice from the adjacent continent is thus present in the middle section of Amery Ice Shelf. The very sharp change in $\delta^{18}O$ values at 270 m appears to mark the bottom of glacier ice; below this depth, the $\delta^{18}O$ values are close to that of ocean water, near 0‰, and the ice is assumed to be frozen sea water. The isotopic values remain near SMOW to the base of the core, at a depth of 315 m, which does not reach the base of the ice shelf, estimated from radar investigations to be at 428 m. If the same ice exists from 270 m to the bottom, this implies a thickness of sea water ice of 158 m. Independently of the oxygen isotope distribution, the salt content of the ice core has been estimated from electrical conductivity measurements. A similar profile is obtained with low conductivities up to 270 m depth, followed by a significant increase in the bottom part of the core. Using measured velocities and distances to the grounding line, an average bottom freezing rate of 30 cm/year is calculated. The S-shaped temperature curve through the Amery Ice Shelf analysed in Section 6.2 gives, by matching it with a freezing rate of 0.4 m/year, a rate close to that obtained above. However, these rates are quite high if, as is believed, it is sea water rather than fresh water which is freezing at the base. The presence of a spike in the isotopic profile (at 260 m) and the lowering of electrical conductivity at the bottom of the core are complications that must be understood before such a freezing rate can be accepted as a reliable value.

Near-surface mean isotopic values for the Ross Ice Shelf are given in Fig. 6.8. The δ values decrease with distance from the sea, and since δ values also decrease to the south and to the west of Roosevelt Island, the main supply of moisture apparently moves over the ice shelf from the north east. Koerner (1979) has advanced an explanation for this effect in the Canadian Arctic, where he believes that the decrease with distance from the source of moisture is a result of isobaric precipitation processes on the flat upper surface of the ice cover, rather than the influence of orographic processes as experienced over rising ice sheets.

The floating tongue of Koettlitz glacier which extends for a distance of approximately 50 km into McMurdo Sound has attracted attention for some time, as fish and other organic remains have been observed on its ablating surface and were considered to be initially incorporated into the bottom of the glacier during freezing-on (Debenham 1920; Swithinbank et al. 1961; Gow et al. 1965).

Gow and Epstein (1972) have used stable isotopes to trace the origin of ice in this floating ice tongue. A schematic cross-section is given in Fig. 6.9 and stable isotope and salinity variation in Table 6.1. Salinities measured in cores from hole 5 along with isotopic values clearly demonstrate a glacial origin for this ice. On the basis of present-day relationships between δ values and temperature or elevation, the ice in hole 5 comes from a region about 40 km upstream on the Koettlitz glacier, at an elevation of about 1000 m. Ice throughout holes 3 and 1 clearly originates from sea water. The δ values are positive and

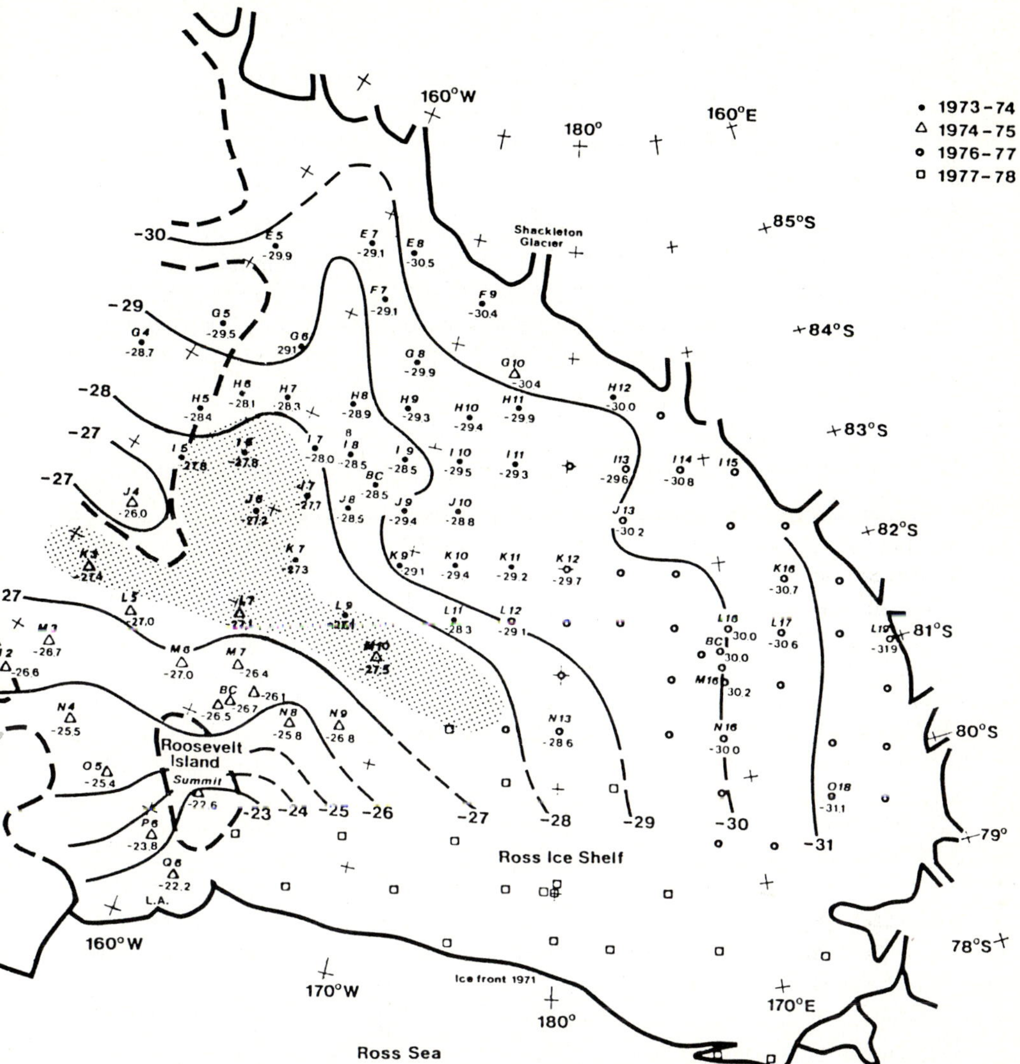

Fig. 6.8. Areal variation of the mean isotopic composition (δ permil) of the upper firn of the Ross Ice Shelf. *Shaded area* coldest 10-m temperatures (less than −28 °C). (Clausen et al. 1979, Fig. 3)

only slightly enriched relative to SMOW, in accordance with the respective equilibrium fractionation coefficients for deuterium and oxygen 18 (1.0208 and 1.003 respectively). This indicates that the lower half of the 50-km Koettlitz ice tongue is composed of sea ice up to 15 m thick. This progressive replacement from ice of pure glacial origin to ice of marine origin downstream is accomplished by the combined processes of ablation of terrestrial ice at the upper surface and freezing-on of sea water at the bottom interface. The up-

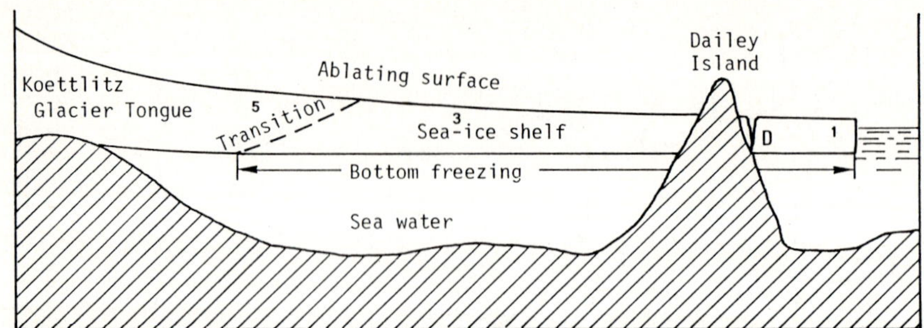

Fig. 6.9. Schematic cross-section of Koettlitz glacier tongue, Antarctica, depicting processes involved in its formation. Numbers *1, 3, 5* refer to holes indicated in Table 6.1. (Gow and Epstein 1972, Fig. 2)

Table 6.1. Stable isotope and salinity variations of the Koettlitz Glacier tongue, McMurdo Sound, Antarctica. (Gow and Epstein 1972, Table 1)

Sample location	Sample depth (m)	δD (‰)	δ^{18}O (‰)	Salinity (‰)	Ice type
Hole 1	0.1	+ 18.3	+ 2.51	0.2	Sea
	1	+ 11.8	+ 1.57	1.00	Sea
	4	+ 15.2	+ 1.67	1.44	Sea
	8	+ 16.1	+ 1.80	2.39	Sea
	11	+ 12.9	+ 1.51	3.13	Sea
	12	+ 14.6	+ 1.37	3.19	Sea
	12.8	+ 15.4	+ 1.61	3.76	Sea
Hole 3	2	+ 14.4	+ 1.76	2.10	Sea
	6	+ 13.8	+ 1.74	3.26	Sea
	9	+ 15.2	+ 1.90	1.75	Sea
	11	+ 13.6	+ 1.66	3.82	Sea
	12	+ 12.8	+ 1.77	2.88	Sea
	12.9	+ 15.1	+ 1.83	3.51	Sea
	13	+ 13.7	+ 1.85	5.26	Sea
	[a]	− 5.6	− 1.12	41.0	Sea
Hole 5	1	− 288.8	− 38.17	0.05	Glacial
	7	− 257.3	− 33.46	0.01	Glacial

For location of samples see Fig. 6.9.
[a] Sea water from bottom.

ward convex temperature distribution in the ice resulting from this bottom freezing has already been described (see Sect. 6.2). Taking into account the ablation rate – around 1.7 m/year – the most probable value for the rate of basal freezing is 1 m/year. This change from a glacial ice tongue to a sea ice shelf allows the incorporation of marine animals and their subsequent appearance at the ice surface. In the vicinity of Dailey Island, near site D on Fig. 6.9, the isotopic composition of the ice is neither indicative of glacial ice nor of sea ice. In fact, this part of the Koettlitz ice tongue is composed of fresh-water ice derived from the meltwater of desalinated sea ice. Gow and Epstein (1972) considered that the water resulting from the melting of old desalinated sea ice drains down tidal cracks to the underside of the ice shelf, where it forms a stable layer of fresh water.

6.5 The Case of the Ward Hunt Ice Shelf

The Ward Hunt Ice Shelf is situated along the northern coast of Ellesmere Island. The ice shelf has an area of 440 km^2, a thickness of about 50 m and is located at the entrance of Disraeli Fiord, where it acts as a floating dam, causing water stratification in the fiord, in the form of a layer of fresh water about 40 m deep overlying sea water (Fig. 6.10). It is now generally accepted that the ice shelf originally grew in situ as a sheet of fast ice and has subsequently thickened by snow accumulation and "meteoric" ice formation. Isotope and salinity measurements of ice cores drilled through the ice shelf are reported by Jeffries et al. (1988).

In the western part of the ice shelf, ice salinity and $\delta^{18}O$ values show considerable variation (Fig. 6.11). Brine upwelling occurred in the borehole to the level indicated in the figure and this is probably the cause of the high ice salinities at the base of the profile, which does not reach the bottom of the shelf. However, $\delta^{18}O$ values are not affected. At a depth of about 10 m, ice with higher salinity is encountered in association with $\delta^{18}O$ values that re-

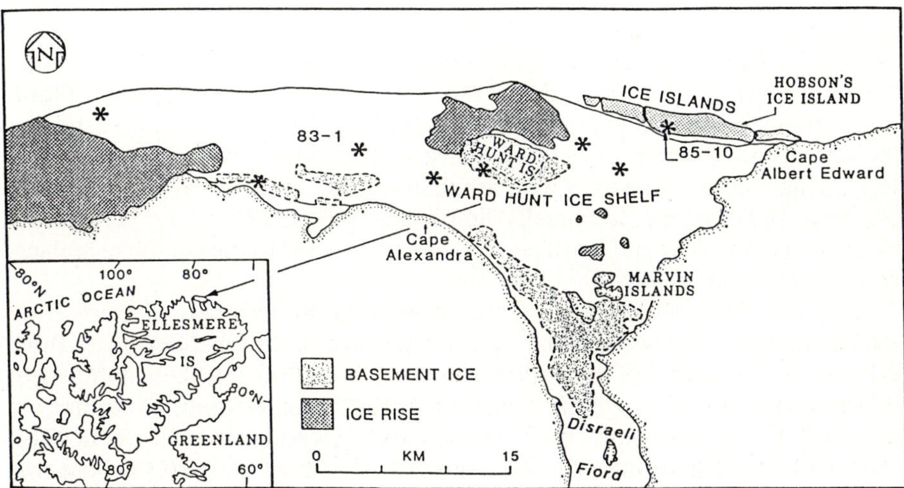

Fig. 6.10. Location map of ice cores (*asterisks*) drilled in Ward Hunt Ice Shelf since 1982. (Jeffries et al. 1988, Fig. 1)

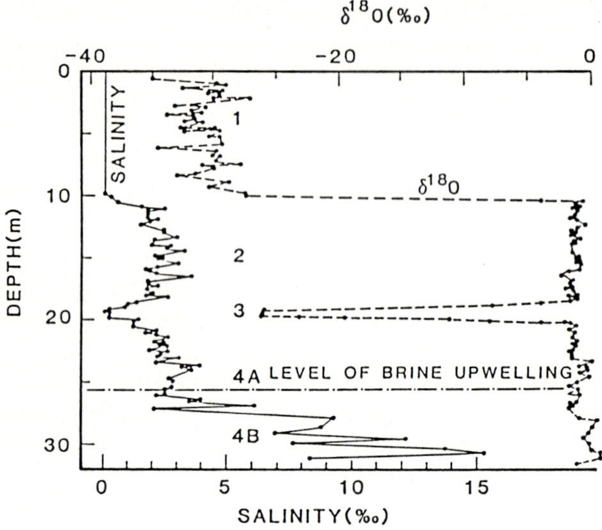

Fig. 6.11. Salinity and $\delta^{18}O$ profiles in ice core 83-1, 9 km west of Ward Hunt Island. Sixty-eight hours after drilling ceased at 31.79 m, brine had upwelled or infiltrated into the borehole to a level of 25.63 m below the ice surface. Ice-core salinity increases sharply just below this level; for this reason stratum 4 has been divided into *A* and *B*. (Jeffries et al. 1988, Fig. 2)

main close to SMOW, indicating ice accretion from seawater below the western ice shelf. This type of accretion is interrupted at a depth of 20 m where non-saline ice and lower $\delta^{18}O$ values are suggestive of ice having formed from fresh water at some past time.

In contrast, in the eastern ice shelf cores (Fig. 6.12) there is an absence of saline ice and $\delta^{18}O$ values fluctuate around -30‰ to the bottom of the shelf. This indicates fresh water ice of meteoric origin. Jeffries et al. (1988) consider the source of the fresh water below the eastern ice shelf to be Disraeli Fiord. The presence of tritium in ice at the bottom of the eastern ice cores means that accretion is quite recent, suggesting a mean accretion rate of 95 mm/year since 1963. This is to be compared with an ablation rate over the same time period of about 100 mm/year (Hattersley-Smith and Serson 1970). The balance between surface ice loss and basal accretion may have led to the complete replacement of ice from below.

Seawater flows into Disraeli Fiord beneath the western part of Ward Hunt Ice Shelf, thereby allowing saline ice to be accreted at the base. Sea water flows out of the fiord beneath the eastern part of Ward Hunt Ice Shelf, but a layer of fresh water exists above the saline sea water. Thus, freezing at the bottom of the eastern part of Ward Hunt Ice Shelf gives rise to fresh water ice, in direct contrast to the situation in the western part. Jeffries et al. (1988) believe that a previous warmer climatic period was responsible for a more widespread occurrence of a fresh water layer below the ice shelf and this may explain the layer with low salinity and low $\delta^{18}O$ values observed within the western ice core.

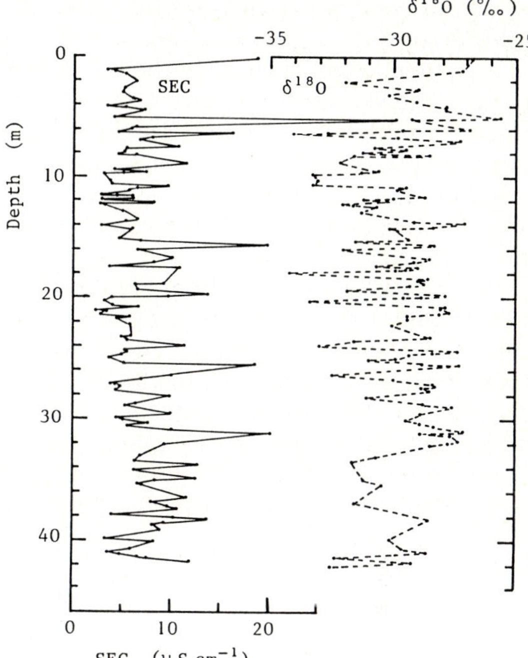

Fig. 6.12. Specific electrolytic conductivity (SEC) and $\delta^{18}O$ profiles in ice core 85 10, "Hobson's Ice Island"/east Ward Hunt Ice Shelf. (Jeffries et al. 1988, Fig. 4)

The Ward Hunt Ice Shelf can be partly considered as a non-saline ice body owing its origin to water freezing at the bottom, the consequence of a peculiar oceanic circulation pattern. In such a case, the growth of the ice shelf is largely independent of supply of glacial origin from an adjacent land mass. It is not known if such an ice body can grow until it becomes grounded and develops into an ice cap as Hughes (1987) implies in his marine ice transgression hypothesis.

6.6 Freezing Rates in the Marine Environment

In Chapter 2 the composition in stable isotopes of ice formed by the progressive migration of a freezing front in water was considered. This is likely to be influenced by the freezing rate, and theory has been developed to deduce freezing rates from the isotopic distribution in ice. This distribution can be determined from a knowledge of the isotopic concentration variations in the water at the freezing interface. Although mixing can be achieved throughout the bulk of the liquid, a zone in which transport takes place by diffusion only exists adjacent to the ice-water interface as a boundary layer of thickness B.L.T. Taking this into account in a box diffusion model using a finite difference approximation of Fick's law of diffusion, Souchez et al. (1987) have been able to predict the isotopic distribution in the ice in the course of freezing for a given freezing rate.

In sea water, the isotopic composition of the water outside the boundary layer is constant during the freezing process because the water layer is more than one order of magnitude thicker than the ice formed. In such circumstances, the isotopic composition of the ice is directly related to the freezing rate for a given B. L. T. value. Since this thickness does not depend on the freezing rate, it is reasonable to assume that it is constant during the freezing process. If the freezing rate changes during the process, a shift will occur towards another isotopic value in the ice in accordance with the new freezing rate. The reader is referred to Section 2.4 for an explanation of these developments.

A first year sea ice core from Breid Bay in East Antarctica has been studied in this context by Souchez et al. (1988a). Tison and Haren (1989) have shown that the evolution of this sea-ice cover was controlled by thermodynamic rather than dynamic processes, a situation different from the one demonstrated by Lange et al. (1989) in the Weddell Sea. Figure 6.13 gives a core profile showing both frazil and congelation ice layers, the δD distribution in the core, the Na concentration distribution and variations in a mean crystal elongation parameter. This core has a length of 1.64 m but only the congelation ice layers, containing elongated crystals, are considered since only in this kind of ice does the process of formation involve the migration of a freezing front through water.

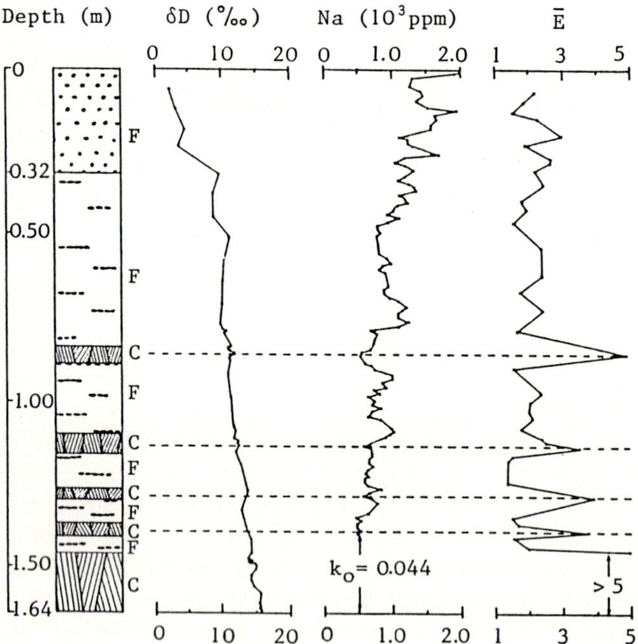

Fig. 6.13. Properties of sea ice in Breid Bay, Antarctica. The mean crystal elongation parameter $\bar{E}$ is the mean ratio between crystal length and width in the vertical plane. k_0 is the solute distribution coefficient at zero growth rate. *F* frazil ice; *C* congelation ice. (Souchez et al. 1988a, Fig. 2)

Figure 6.13 shows that there is a general trend towards higher δD values as one proceeds down through the core. Using an equilibrium fractionation coefficient for deuterium of 1.0208, the δD value at the very bottom corresponds closely to the maximum possible shift:

$$\delta D_{ice} - \delta D_{water} = 15.9‰ - (-4.8‰) = 20.7‰$$

instead of 20.8‰, implying that the growth rate was nearly zero at the bottom of the core. There is, by contrast, a trend towards lower Na concentration as one proceeds from the top to the bottom of the ice core. These trends are understandable if the growth rate diminishes downwards, since solutes are more efficiently rejected from the ice at lower freezing rates, while the heavy isotopes are better incorporated.

To be able to deduce the growth rate from the δD value of a congelation ice layer, the boundary layer thickness must be known. Using two equations derived from the core data, one for δD and the other for Na concentration, Souchez et al. (1988a) were able to estimate the B.L.T. value to be 0.09 cm. Table 6.2 gives the deduced freezing rates for the different congelation ice layers of the profile. Isotopic determination of the freezing rate is likely to be more accurate than the one based on the salinity profile, since it is less affected by brine entrapment and brine drainage mechanisms.

The composition of basal ice accreted at the sole of an ice shelf can itself give some idea of the freezing rate. Table 6.1 gives the salinity and the stable isotope composition of the hole 3 ice core from the Koettlitz ice tongue, together with values of bottom sea water. In this case there is no means of calculating the boundary layer thickness and so the freezing rate cannot be determined as in the analysis of the Breid Bay core described above. However, if the lowest δD value of the core (+12.8‰) is compared to the δD value of sea water underneath (−5.6‰), an apparent fractionation coefficient can be calculated:

$$\alpha_{app} = \frac{1000 + \delta_{ice}}{1000 + \delta_{water}} = \frac{1000 + 12.8}{1000 - 5.6} = 1.0185 \ .$$

Table 6.2. δD values and growth rates for four samples of fine-grained congelation ice (depths: 0.86, 1.14, 1.28, 1.40 m) and three samples of coarse-grained congelation ice (depths: 1.52, 1.57, 1.61 m), Breid Bay, Antarctica. (Souchez et al. 1988a)

Depth (m)	δD (‰)	Growth rate (cm/day)
0.86	11.8	2.29
1.14	12.4	1.92
1.28	13.7	1.16
1.40	13.9	1.05
1.52	14.2	0.89
1.57	14.6	0.67
1.61	15.6	0.15

This shows a significant difference with the equilibrium fractionation coefficient 1.0208, which would have been obtained at a very slow freezing rate. A freezing rate effect is probably indicated in the isotopic composition although it cannot be estimated in this case with the available data. However, the potential of the technique will diminish with the freezing rate as one approaches a situation of equilibrium fractionation. It is likely that such a situation occurs at the low freezing rates normally encountered at the base of ice shelves.

At Camp J9 in the central zone of the Ross Ice Shelf, the rate of bottom freezing over a long time period has been estimated from the thickness of the sea-ice layer at the base (about 6 m) and the rate of movement. Given a distance of about 200 km parallel to the flow lines to the grounding line and a mean longitudinal flow velocity of 300 to 400 m/year, Zotikov (1986) estimates that the accretion of 6 m of sea ice at the base in this area took from 400 to 600 years. The mean annual freezing rate at the bottom is thus 1 to 1.3 cm/year. The situation at Camp J9 seems to be typical of the central inner zone of the Ross Ice Shelf. Temperature measurements were carried out in the water under the ice shelf (at J9). Two definite water layers are present, one, about 80 m thick, with a temperature very close to the freezing point and the other, extending to the end of the recorded temperature profile, with a temperature about 0.5 °C above the freezing point of sea water. Different temperatures have been obtained at a few days' interval, indicating that the water tem-

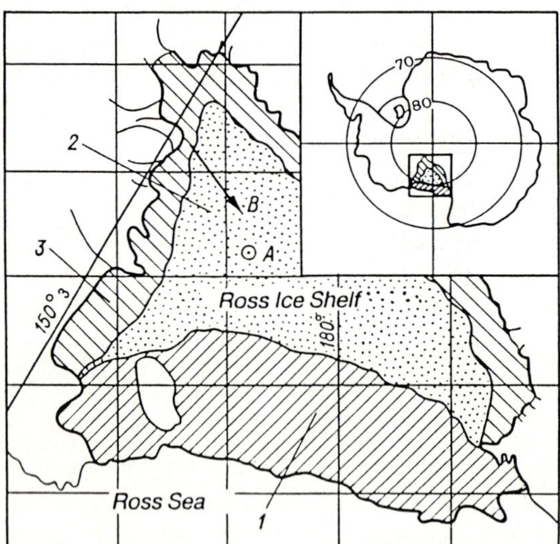

Fig. 6.14. Map of the Ross Ice Shelf. **A** Site of drilling (camp J-9); **B** direction of glacier movement. *1* Melting region; *2* region of bottom freezing caused by upward heat transport through shelf; *3* region of bottom freezing enhanced by arrival of fresh water under the shelf. (Zotikov 1986, Fig. 9.9)

perature beneath the ice shelf varies with time. No supercooled water seems to exist, and heat exchange between the shelf sole and the sea seems to be absent. Bottom freezing therefore occurs at a rate governed by heat conduction through the ice. The inner central zone of slow bottom freezing gives way to a zone of bottom melting in the outer part of the shelf (Fig. 6.14). A zone of enhanced bottom freezing exists along the edge of the ice shelf near the grounding line, a product of the presence of meltwater flowing at the base of the continental glaciers feeding the ice shelf. This water is discharged into the sea at the grounding line and forms a low salinity water layer and higher freezing rates are encountered in this zone, reaching a few tens of cm/year. The effect of a fresh water layer at the surface of sea water is to enhance the basal freezing rate. In the next section, consequences of this fact in terms of the glacier supply to the ocean will be considered.

6.7 The Glacial Supply to the Ocean

Surface sea waters in polar regions near glaciated areas are partly derived from glacial melt. These waters are easily recognizable by their low salinities and by their very negative isotopic δ values. As seen in Section 6.4, this is due to the very negative $\delta^{18}O$ or δD values of glacier ice compared to oceanic water which has an isotopic composition close to SMOW. Glacial melt not only supplies fresh water to the ocean but also mineral particles embedded in the ice or transported by meltwater. Freezing-on can also occur, and this represents a further complicating factor. Indeed, the contact zone between glacier and ocean is characterized by complex phase interactions.

The relatively fresh water derived from the melting of sea ice differs in its deuterium or oxygen 18 content from that derived from the melting of snow and glacier ice. This is a consequence of the fact that the freezing of sea water produces ice with a deuterium and oxygen 18 content close to that of SMOW, whereas snow precipitated in the polar regions has very negative isotopic values owing to progressive fractionation during evaporation and subsequent precipitation during its passage polewards through the atmosphere.

By examining the relationship between deuterium or oxygen 18 concentration and salinity in the subsurface waters of the polar seas, it is possible to identify the sources of the fresh water diluent. Meltwater from sea ice may be expected to reduce the salinity of the surface sea water without changing greatly its deuterium or oxygen 18 content. In Fig. 6.15 from Redfield and Friedman (1969), a comparison is given between the δD and salinity of waters from the Canadian Basin of the Arctic Ocean (Fig. 6.15 A) and from Tanquary fiord in Ellesmere Island (Fig. 6.15 B). Surface waters in A show salinity values (S) from 25 to 31‰ without any significant change in δD, indicating the dominant contribution of sea ice melting. Below 50 m, samples of water are aligned on a $\delta D/S$ slope of 7. By extrapolation, a value of $\delta D = -242$‰ is obtained for zero salinity. This value indicates that the diluent must be meltwater from the Greenland Ice sheet. In contrast, the $\delta D/S$ slope for surface water samples

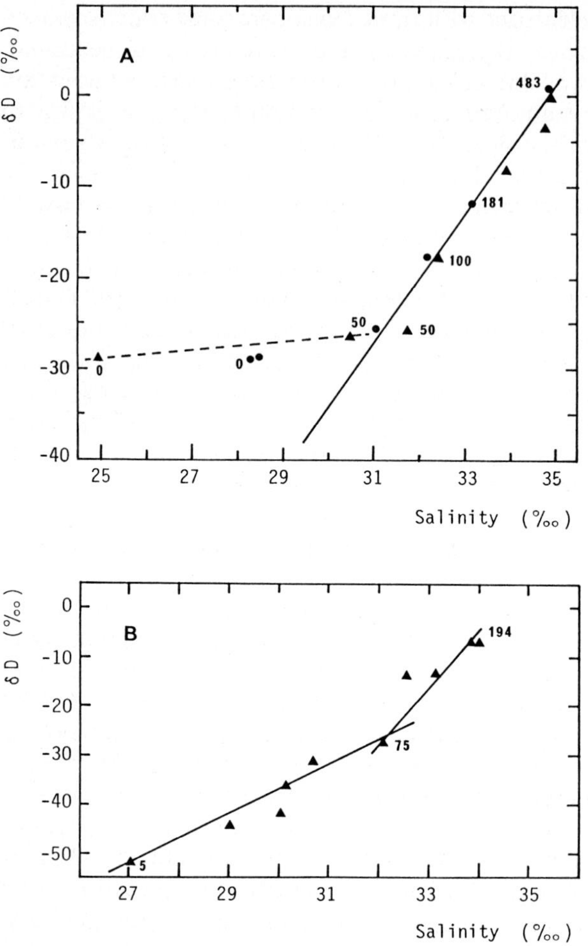

Fig. 6.15. A Deuterium-salinity relation in waters of the Canadian Basin of the Arctic Ocean. Based on samples from Canadian icebreaker Labrador. *Numbers beside symbols* indicate depth in meters. **B** Deuterium-salinity relations in waters of Tanquary Fiord, Ellesmere Island, Arctic Canada. *Numbers beside symbols* indicate depth in meters. (After Figs. 3 and 8 in Redfield and Friedman 1969)

at Tanquary fiord (Fig. 6.15B) is about 5, leading to a value of the apparent diluent of about $-180‰$. Assuming that the δD of sea ice is $-10‰$ and that of glacier ice $-250‰$, the diluent is a mixture of 29% meltwater from sea ice and 71% meltwater from glacier ice. Thus, in coastal areas, the greater part of the diluent is derived from glacier ice while, in the open sea, the surface layers are diluted predominantly by meltwater from sea ice.

Paren and Potter (1984) have shown that there is a source of error implicit in the use of δ-salinity diagrams. Due to the way salinity is defined, it is physically more meaningful to use $(1-S)\delta$-salinity diagrams rather than δ-salinity diagrams. Such a procedure introduces no significant error during extrapolation. Studying the $(1-S)\delta^{18}O$-salinity diagram of the waters beneath the northern ice front of George VI Ice Shelf in the Antarctic Peninsula, Potter et al. (1984) indicate that the diluent has a $\delta^{18}O$ value of −20.3‰. This is in good accordance with the $\delta^{18}O$ value of −20.8‰ representing the mean δ value of the catchment furnishing ice to the George VI Ice Shelf. Glaciological and oceanographic evidence point toward bottom melting of the shelf as the likely source of diluted water. Assuming a steady state, the basal melt rate is estimated to be about 2 m/year. However, in the central region of the ice shelf near Palmer Land, a layer of low salinity water about 70 m thick has been detected immediately beneath the ice shelf. This water is close to its freezing point and is underlain by colder, saline water. The presence of this colder and more saline water leads to the in situ freezing of the fresher water lying above it. The frazil ice is accumulated at the base of George VI Ice Shelf in this region. The origin of the low salinity water is the drainage of surface meltwater through the ice shelf in areas where it is thin and crevassed as a result of the coastal topography. The existence of low salinity waters depleted in heavy isotopes as a result of ice melting seems therefore to play a significant role in bottom freezing here, as in other places.

Freezing at the base of tidewater glaciers near the grounding line can constitute an efficient mechanism of erosion since the debris-rich basal zone thereby produced is displaced downglacier to the edge of the floating tongue. Crary (1966) considers that the decrease in ice thickness down the inlet as a result of the influence of shear stresses on the side walls, coupled with such a freezing-on process, explains the development of fiords with a distinctive threshold near the outlet and a deep basin inland. This does not dismiss the current view that many fiords are primarily tectonic features. The movement of material by floating ice could be likened to an endless belt or chain saw; the morphology of the floor must, to a large extent, be determined by the configuration of the lower boundary of the ice, and multiple basins could result from different boundary conditions.

Even if freezing occurs near the grounding line, the inevitable fate of the particles embedded in the ice is a release by melt out into the ocean. Concerning this process, the thermal properties of basal debris themselves have an effect on the melt rate. As indicated by Drewry (1986), if the coefficient of thermal conductivity of debris-rich basal ice is higher than that of pure ice, less basal melt occurs since the heat flux from the sea water can be easily conducted through the ice. On the other hand, accelerated melting takes place in the reverse situation. In turn, the coefficient of thermal conductivity is largely dependent on the concentration and size of particles embedded in the ice.

If at the grounding line, ice melts as the glacier begins to float, then the entrained sediments of the stratified facies, which usually exists close to the sole, are released. Further out from the grounding line, the rate of sedimentation of material originally entrained in stratified basal ice will fall off rapidly as melting proceeds. This is due to the decrease in the vertical concentration of debris in ice upwards from the sole and finer fractions from the upper dispersed ice facies as opposed to the lower stratified facies, increasingly dominate deposition. Drewry and Cooper (1981) consider that if there is vigorous melting close to the grounding line, all included debris may melt out within the first few tens of kilometres because of the relatively thin vertical extent of the debris-rich basal zone. Thus, the basal zone is only preserved far from the grounding line in situations characterized by the near absence of basal melting. The discharge of meltwater from the glacier base into the ocean or the penetration of surface melt through the floating ice is of paramount importance in this context.

Conclusion: Ice Composition, Glacier Dynamics and Global Changes

The increase in concentration of greenhouse gases in the atmosphere – carbon dioxide, CFCs, water vapour and methane – is likely to give rise to climatic changes. These gases have absorption properties in the infra-red portion of the radiation spectrum and, by a greenhouse effect, control to some extent the surface temperature of the earth. Concentration changes in the atmosphere due to human activities are thought to produce a general warming of the lower troposphere, particularly in the polar regions. Questions thus arise concerning the relationship between such a temperature increase and the behaviour of polar ice sheets and ice caps and, in turn, about its influence on global sea level. Can the study of ice composition give some clue about the dynamics of ice masses and thus contribute to the understanding of the global changes that may be induced by this climatic warming? The first part of this question has been answered at length in this book and we will now briefly concentrate on the second part.

Different approaches to the problem can be considered. First, the palaeoclimatic record can be studied. As an example, there exist in the ice record of polar ice sheets and ice caps relatively short-term events which are clearly identified as compositional spikes. Such short-term events, giving spikes in the ice record, are worth considering, since, through a general study of the different evidence relating to environmental conditions data at a particular time period, relationships can be established which allow an appraisal of the way in which the global atmosphere-ocean-cryosphere system reacts to a known perturbation.

Another approach is to consider how ice composition is acquired. A study at specific interfaces is clearly needed in this case in order to evaluate boundary conditions. What are the factors involved in the acquisition of the ice characteristics during transformation of snow into ice at the atmosphere-glacier interface? What are the environmental controls on ice accretion at the base of ice sheets and ice caps along the ice-bed interface? What conditions give rise to bottom freezing at the sole of ice shelves, at the sea water-glacier ice interface? With such information, it is possible to understand glacier dynamics better. Now, monitoring ice composition changes due to climatic warming allows us to estimate the kind of response expected from the ice masses. For example, an increase in percolation of surface meltwater changes the isotopic composition of the firn and finally of the ice. More water penetrating the snow cover gives rise to more refreezing at depth. This refreezing of percolating meltwaters is responsible for a warming of the ice sheet, which has an influence on flow.

Similarly, freezing of subglacial meltwaters at the glacier sole helps to maintain the glacier base at the pressure-melting point by the latent heat liberated during ice formation. The sole of an ice sheet needs to be at the pressure-melting point for substantial sliding to occur. As such, the study of ice composition provides an understanding of basal processes which are critical to the flow of ice masses as they represent the lower boundary conditions.

Among the glaciological problems raised by a mean temperature increase in the polar regions is the stability of the Antarctic ice shelves. Ice shelves are vulnerable to both atmospheric and oceanic warming. In the Antarctic Peninsula, the 0 °C isotherm for the warmest month is almost parallel to, but a short distance outside, the northern limit of ice shelves (Robin 1979). The main ice shelves would probably recede southward if summer temperatures rose 5 °C at a latitude 80°S, a temperature increase implied by the majority of climatic models for a doubled CO_2 increase in concentration. Temperate ice shelves do not seem to exist in nature at present, but ice would become temperate if rising air temperatures produced enough meltwater to percolate downwards and destroy the previous winter's cold wave (Mercer 1978). Ice shelves will melt at the base if the water they are in contact with is above its freezing point, as indicated by the fact that they are absent from the coasts of the Antarctic Peninsula, where sea temperatures rise above -1.5 °C during the warmest month.

The West Antarctic Ice Sheet, which is grounded far below sea level, is buttressed by ice shelves which are highly sensitive to climatic changes. Most of the ice drainage from the ice sheet is accomplished by ice streams. What is the influence of ice-shelf weakening on ice-stream velocity? The ice stream pushes the ice shelf seaward past its margins and around the grounded ice rises and there is a corresponding compressive force at the grounding line between ice stream and ice shelf (Thomas 1985). The results obtained by Thomas reveal a sensitive response of ice-discharge rates to ice shelf thinning. Collapse of part of the West Antarctic Ice Sheet through accelerating ice streams is a possibility. This could give a 3-m increase in global sea level in about 500 years. However, the data on bottom freezing given by Zotikov (1986) indicate an enhanced freezing rate when fresh water exists beneath the ice shelf. The accelerated movement of ice streams and outlet glaciers would produce more freshwater and the resultant freezing when this water is introduced beneath the ice shelf at the grounding line will result in the local thickening of the ice shelf. Thus, the grounding line may move downstream and this will increase the resistance of the ice shelf. As a result, the ice movement will be slowed down and a stabilization of the ice sheet-ice stream-ice shelf system occurs. These ideas call for a reconsideration of the reaction of ice shelves to climatic changes, as suggested by Zotikov (1986). Obviously, a more intimate knowledge of the system must be gained before valid conclusions can be reached. More ice shelf drilling with the analysis of basal ice to detect bottom freezing and estimate accretion rates is needed for a better understanding of the complexity of the situation.

Ice composition studies involving the analyses of isotopes and impurities in ice cores have given one of the best palaeoclimatic and environmental records available for the last 160000 years, the age of the ice at 2200 m in the Vostok ice core (Jouzel et al. 1987; Barnola et al. 1987). Additionally, ice composition studies also represent an effective tool in our understanding of glacier dynamics. The authors hope that this book has convinced the reader that such a subject can be exciting.

References

Aharon P (1988) Oxygen, carbon and U-series isotopes of aragonites from Vestfold Hills, Antarctica: clues to geochemical processes in subglacial environments. Geochim Cosmochim Acta 52:2321–2331

Alley RB (1988) Fabrics in polar ice sheets: development and prediction. Science 240:493–495

Alley RB (1989) Water-pressure coupling of sliding and bed deformation: I Water system. J Glaciol 35(119):108–118

Alley RB, Blankenship DD, Bentley CR, Rooney ST (1986) Deformation of till beneath ice stream B, West Antarctica. Nature 322:57–59

Ambach W, Dansgaard W, Eisner H, Möller J (1968) The altitude effect on the isotopic composition of precipitation and glacier ice in the Alps. Tellus 20:595–600

Ambach W, Eisner H, Url M (1973) Seasonal variations in the tritium activity of run-off from an Alpine glacier, Kesselwandferner (Oetztal Alps). Proceedings of the symposium on the hydrology of glaciers, Cambridge, September 1969. I.A.H.S. Publication 95, pp 199–204

Ambach W, Eisner H, Elsässer M, Löschhorn U, Moser H, Rauert W, Stichler W (1976) Deuterium, tritium and gross beta activity investigations on Alpine glaciers (Oetztal Alps). J Glaciol 17:383–400

Anderson DM (1967) The interface between ice and silicate surfaces. J Colloid Interface Sci 25(2):174–191

Anderson DM, Morgenstern NR (1973) Physics, chemistry and mechanics of frozen ground: a review. In: Permafrost, the North American contribution to the 2nd International Conference. National Academy of Sciences, Washington DC, pp 257–288

Anderton PW (1974) Ice-fabrics and petrography, Meserve Glacier, Antarctica. J Glaciol 13(68):285–306

Andrews JT (1971) Englacial debris in glaciers. J Glaciol 10(60):410 (Letter)

Andrews JT (1972) Englacial debris in glaciers. J Glaciol 11(61):155 (Letter)

Armstrong TE, Roberts B, Swithinbank C (1973) Illustrated glossary of snow and ice, 2nd edition. Scott Polar Research Institute, Cambridge, 60 p

Arnason B (1969a) The exchange of hydrogen isotopes between ice and water in temperate glaciers. Earth Planet Sci Lett 6(6):423–430

Arnason B (1969b) Equilibrium constant for the fractionation of deuterium between ice and water. J Phys Chem 73(10):3491–3494

Arnason B (1981) Ice and snow hydrology. In: Gat JR, Gonfiantini R (eds) Stable isotope hydrology, deuterium and oxygen-18 in the water cycle. Technical Series (210), International Atomic Energy Agency Vienna, pp 143–175

Barnola JM, Raynaud D, Korotkevich YS, Lorius C (1987) Vostok ice core provides 160.000-year record of atmospheric CO_2. Nature 329(6138):409–414

Bauer VF (1961) Kalkabsätze unter Kalkalpengletschern und ihre Bedeutung für die Altersbestimmung heute gletscherfrei werdender Karrenformen. Z Gletscherkd Glazialgeol 4(3):215–225

Benson CS (1962) Stratigraphic studies in the snow and firn of the Greenland ice sheet. SIPRE Rep 70, 93 p

Berner W, Stauffer B, Oeschger H (1978) Dynamic glacier flow model and the production of internal meltwater. Z Gletscherkd Glazialgeol 13:209–217

Berry FAF (1969) Relative factors influencing membrane filtration effects in geologic environments. Chem Geol 4(1–2):295–301

Bindschadler R (1983) The importance of pressurized subglacial water in separation and sliding at the glacier bed. J Glaciol 29(101):3–19
Blankenship DD, Bentley CR, Rooney ST, Alley RB (1986) Seismic measurements revealed a saturated porous layer beneath an active antarctic ice stream. Nature 322:54–57
Boulton GS (1970) On the origin and transport of englacial debris in Svålbard glaciers. J Glaciol 9(56):213–229
Boulton GS (1972) The role of thermal regime in glacial sedimentation. In: Price R, Sugden D (eds) Polar geomorphology. Inst Br Geogr Spec Publ 4:1–19
Boulton GS (1975) Processes and patterns of subglacial sedimentation: a theoretical approach. In: Wright A, Moseley F (eds) Ices ages: ancient and modern. Seel House, Liverpool pp 7–42
Boulton GS (1978) Boulder shapes and grain-size distributions of debris as indicators of transport paths through a glacier and till genesis. Sedimentology 25:773–799
Boulton GS (1979) Processes of glacier erosion on different substrata. J Glaciol 23:15–38
Boulton GS, Hindmarsh RC (1987) Sediment deformation beneath glaciers: rheology and geological consequences. J Geophys Res 92(B9):9059–9082
Boulton GS, Spring U (1986) Isotopic fractionation at the base of polar and sub-polar glaciers. J Glaciol 32(112):475–485
Buason T (1972) Equation of isotope fractionation between ice and water in a melting snow column with continuous rain and percolation. J Glaciol 11(63):387–405
Budd WF (1972) The development of crystal orientation fabrics in moving ice. Z Gletscherkd Glazialgeol 8(1–2):65–105
Budd WF, Smith IN (1981) The growth and retreat of ice sheets in response to orbital radiation changes. Sea level, ice and climatic change. Proceedings of the Canberra Symposium, IAHS Publication 131, pp 369–409
Burton JA, Prim RC, Slichter WP (1953) The distribution of solute in crystal growth from the melt. Part 1: theoretical. J Chem Phys 21(1):1987–1991
Clapperton C (1975) The debris content of surging glaciers in Svålbard and Iceland. J Glaciol 14(72):395–406
Clarke GKC (1976) Thermal regulation of glacier surging. J Glaciol 16(74):231–250
Clausen HB, Stauffer B (1988) Analyses of two ice cores drilled at the ice-sheet margin in West Greenland. Ann Glaciol 10:23–27
Clausen HB, Dansgaard W, Nielsen JO, Clough JW (1979) Surface accumulation on Ross Ice Shelf. Antarctic J US 5:68–72
Clayton RN, Jones BF, Berner RA (1968) Isotopic studies of dolomite formation under sedimentary conditions. Geochim Cosmochim Acta 32:415–432
Craig H (1961) Isotopic variations in meteoric waters. Science 133:1702–1703
Crary AP (1966) Mechanism for fiord formation indicated by studies of an ice-covered inlet. Geol Soc Am Bull 77:911–930
Crozaz G, Langway CC (1966) Dating Greenland firn-ice cores with Pb-210. Earth Planet Sci Lett 1:194–196
Crozaz G, Picciotto E, De Breuck W (1964) Antarctic snow chronology with Pb-210. J Geophys Res 69(12):2597–2604
Dahl-Jensen D, Gundestrup NS (1987) Constitutive properties of ice at Dye 3, Greenland. The physical basis of ice sheet modelling. Proceedings of the Vancouver Symposium, August, 1987. IAHS Publication 170, pp 31–43
Dansgaard W (1964) Stable isotopes in precipitation. Tellus 16:436–468
Dansgaard W, Johnsen SJ, Clausen HB, Langway CC (1971) Climatic record revealed by Camp Century ice core. In: Turekian KK (ed) Late Cenozoïc glacial ages. Yale University Press, New Haven, Connecticut, pp 37–56
Dansgaard W, Johnsen SJ, Clausen HB, Gundestrup N (1973) Stable isotope glaciology. Medd Grønl 197(2):1–53
Dansgaard W, Clausen HB, Gundestrup N, Hammer U, Johnsen SF, Kristinsdottir PM, Reeh N (1982) A new Greenland deep ice core. Science 218(4579):1273–1277
Debenham F (1920) A new mode of transportation by ice: the raised marine muds of South Victoria Land, Antarctica. Q J Geol Soc Lond 75:51–76

Deutsch S, Ambach W, Eisner H (1966) Oxygen isotope study of snow and firn on an Alpine glacier. Earth Planet Sci Lett 1:197–201

Doake CSM (1976) Thermodynamics of the interaction between ice shelves and the sea. Polar Rec 18(112):37–41

Dreimanis A, Vagners UJ (1971) Bimodal distribution of rock and mineral fragments in basal till. In: Goldthwait RE (ed) Till, a symposium. Ohio State University Press, Columbus, pp 237–250

Drewry D (1983) Antarctic ice sheet: aspects of current configuration and flow. In: Gardner R, Scoging H (eds) Megageomorphology. Clarendon Press, Oxford, pp 18–38

Drewry DJ (1986) Glacial geologic processes. Arnold, London, 276 p

Drewry DJ, Cooper APR (1981) Processes and models of Antarctic glaciomarine sedimentation. Ann Glaciol 2:117–122

Duval P, Lorius C (1980) Crystal size and climatic record down to the last ice age from Antarctic ice. Earth Planet Sci Lett 48:59–64

Echelmeyer K, Wang Zhongxiang (1987) Direct observation of basal sliding and deformation of basal drift at sub-freezing temperatures. J Glaciol 33(113):83–98

Epstein B (1947) The mathematical description of certain breakage mechanisms leading to the logarithmico-normal distribution. J Franklin Inst 244:471–477

Epstein S, Sharp RP, Goddard I (1963) Oxygen isotope ratios in Antarctic snow, firn and ice. J Geol 71:698–720

Faure G (1977) Principles of isotope geology. Wiley, New York, 464 p

Fischer DA, Koerner RM (1986) On the special rheological properties of ancient microparticle laden Northern Hemisphere ice as derived from bore-hole and core measurements. J Glaciol 32(112):501–510

Ford DC, Fuller PG, Drake JJ (1970) Calcite precipitates at the soles of temperate glaciers. Nature 226:441–442

French HM (1976) The periglacial environment. Longman, London, 309p

French HM, Harry DG (1988) Nature and origin of ground ice, Sandhills moraine, Southwest Banks Island, Western Canadian Arctic. J Quaternary Sci 3(1):19–30

Friedman I, Redfield AC, Schoen B, Harris J (1964) The variation of the deuterium content of natural waters in the hydrologic cycle. Rev Geophys 2(1):177–189

Gat RT (1981) Properties of the isotopic species of water: 'the isotopic effect'. In: Gat JR, Gonfiantini R (eds) Stable isotope hydrology, deuterium and oxygen-18 in the water cycle. Technical Series (210), International Atomic Energy Agency Vienna, pp 7–18

Glen JW (1955) The creep of polycrystalline ice. Proc R Soc Lond Ser A 228:519–538

Glen JW, Homer DR, Paren JG (1977) Water at grain boundaries: its role in the purification of temperate glacier ice. Isotopes and impurities in snow and ice. Proceedings of the Grenoble Symposium August/September 1975. IAHS Publication 118, pp 263–271

Gordon JE, Darling WG, Whalley WB, Gellatly AF (1988) δD-$\delta^{18}O$ relationships and the thermal history of basal ice near the margins of two glaciers in Lyngen, North Norway. J Glaciol 34(118):265–268

Gow AJ (1970) Preliminary results of studies of ice cores from the 2164 m deep drill hole, Byrd Station, Antarctica. Glaciological exploration. Proceedings of the Hanover Symposium 1985. IAHS Publication 86, pp 78–90

Gow AJ, Epstein S (1972) On the use of stable isotopes to trace the origins of ice in a floating ice tongue. J Geophys Res 77(33):6552–6557

Gow AJ, Williamson T (1975) Gas inclusions in the Antarctic ice sheet and their glaciological significance. J Geophys Res 80:5101–5108

Gow AJ, Williamson T (1976) Rheological implications of the internal structure and crystal fabrics of the West Antarctic ice sheet as revealed by deep core drilling at Byrd Station. Geol Soc Am, Bull 87:1665–1677

Gow AJ, Weeks WF, Hendrickson G, Rowland R (1965) New light on the mode of uplift of the fish and fossiliferous moraines of the McMurdo Ice Shelf, Antarctica. J Glaciol 5:813–828

Gow AJ, Ueda HT, Garfield DE (1968) Antarctic ice sheet: preliminary results of first core hole to bedrock. Science 161(3845):1011–1013

Gow AJ, Epstein S, Sheehy W (1979) On the origin of stratified debris in ice cores from the bottom of the Antarctic Ice Sheet. J Glaciol 23(89):185–192

Haeberli W (1975) Eistemperaturen in den Alpen. Z Gletscherkd Glazialgeol 11(2):203–220

Haeberli W (1985) Creep of mountain permafrost: internal structure and flow of Alpine rock glaciers. Mitt Versuchsanstalt Wasserbau, Hydrologie Glaziol Zürich 77:142

Haefeli R, Von Sury H (1975) Strain and stress in snow, firn and ice along the EGIG profile of the Greenland ice sheet. Snow mechanics. Proceedings of the Grindelwald Symposium, April 1974. IAHS Publication 114, pp 342–352

Hageman R, Lohez P (1978) Twin mass spectrometers for simultaneous isotopic analysis of hydrogen and oxygen in water. Adv Mass Spectrometry 7:504–508

Haldorsen S (1981) Grain-size distribution of subglacial till and its relation to glacial crushing and abrasion. Boreas 10:91–105

Hallet B (1975) Subglacial silica deposits. Nature 254:682–683

Hallet B (1976) Deposits formed by subglacial precipitation of $CaCO_3$. Geol Soc Am Bull 87:1003–1015

Hallet B, Lorrain R, Souchez R (1978) The composition of basal ice from a glacier sliding over limestones. Geol Soc Am Bull 89(2):314–320

Hambrey MJ (1975) The origin of foliation in glaciers: evidence from some Norwegian examples. J Glaciol 14:181–185

Hambrey MJ, Müller F (1978) Structures and ice deformation in the White Glacier, Axel Heiberg Island, Northwest Territories, Canada. J Glaciol 20:41–66

Hammer CU, Clausen HB, Dansgaard W, Gundestrup N, Johnsen SJ, Reeh N (1978) Dating of Greenland ice cores by flow models, isotopes, volcanic debris and continental dust. J Glaciol 20:3–26

Hanshaw BB, Hallet B (1978) Oxygen isotope composition of subglacially precipitated calcite: possible paleoclimatic implications. Science 200:1267–1270

Hattersley-Smith G, Serson H (1970) Mass balance of Ward Hunt Ice Rise and Ice Shelf: a 10-year record. J Glaciol 9:247–252

Helfferich F (1962) Ion exhange. McGraw Hill, New York, 624 p

Herron S, Langway C (1979) The debris-laden ice at the bottom of the Greenland ice-sheet. J Glaciol 23(89):193–207

Hobbs PV (1974) Ice physics. Clarendon Oxford, 837 p

Hodgson DR, Vincent JS (1984) A 10000 yr BP extensive ice shelf over Viscount Melville Sound, Arctic Canada. Quaternary Res 22:18–30

Hoekstra P, Chamberlain E (1964) Electro-osmosis in frozen soils. Nature 203:1406–1407

Hooke RLB (1970) Morphology of the ice-sheet margin near Thule, Greenland. J Glaciol 9(57):303–324

Hooke RLB (1973 a) Structure and flow in the margin of the Barnes Ice Cap, Baffin Island, N.W.T., Canada. J Glaciol 12(66):423–438

Hooke RLB (1973 b) Flow near the margin of the Barnes Ice Cap, and the development of ice-cored moraines. Geol Soc Am Bull 84:3929–3948

Hooke RLB (1976) Pleistocene ice at the basal of the Barnes Ice Cap, Baffin Island, N.W.T., Canada. J Glaciol 17(75):49–60

Hooke RLB (1977) Basal temperatures in polar ice sheets: a qualitative review. Quaternary Res 7:1–13

Hooke RLB, Clausen HB (1982) Wisconsin and Holocene $\delta^{18}O$ variations, Barnes Ice Cap, Canada. Geol Soc Am Bull 93(8):784–789

Hooke RLB, Hudleston PJ (1978) Origin of foliation in glaciers. J Glaciol 20:285–299

Hooke RLB, Dahlin BB, Kauper MT (1972) Creep of ice containing dispersed fine sands. J Glaciol 11(63):327–336

Hooke RLB, Gao XQ, Jacka TH, Souchez RA (1988) Rheological contrast between Pleistocene and Holocene ice in Barnes Ice Cap, Baffin Island, N.W.T., Canada: a new interpretation. J Glaciol 34(118):364–365

Houghton G, Ritchie PD, Thomson JA (1962) The rate of solution of small stationary bubbles and the diffusion coefficients of gases in liquid. Chem Eng Sci 17:221–227

Hudleston PJ (1976) Recumbent folding in the base of the Barnes Ice Cap, Baffin Island, Northwest Territories, Canada. Geol Soc Am Bull 87(12):1684–1692

Hudleston PJ (1977) Progressive deformation and development of fabric across zones of shear in glacial ice. In: Saxena SK, Bhatta-Charjis (eds) Energetics of Geological Processes. Springer, Berlin Heidelberg New York, pp 121–150

Hudleston PJ, Hooke RLB (1980) Cumulative deformation in the Barnes Ice Cap and implications for the development of foliation. Tectonophysics 66:127–146

Hughes T (1973) Glacial permafrost and pleistocene ice ages. In: Permafrost, the North American contribution to the 2nd International Conference. National Academy of Science, Washington DC, pp 213–223

Hughes TJ (1987) The marine ice transgression hypothesis. Geogr Ann 69A(2):237–250

Iken A, Flotron A, Haeberli W, Röthlisberger H (1979) The uplift of Unteraargletscher at the beginning of the meltseason, a consequence of water storage at the bed ? J Glaciol 23(89):430–432

Jeffries MO, Sachinger WM, Krouse HR, Serson HV (1988) Water circulation and ice accretion beneath Ward Hunt ice shelf (northern Ellesmere Island, Canada), deduced from salinity and isotope analysis of ice cores. Ann Glaciol 10:68–72

Johnsen SJ (1977) Stable isotope homogenization of polar firn and ice. Isotopes and impurities in snow and ice. Proceedings of the Grenoble Symposium August/September 1975, IAHS Publication 118, pp 210–219

Johnsen SJ, Dansgaard W, Clausen HB, Langway C (1972) Oxygen isotope profiles through the Antarctic and Greenland ice sheets. Nature 235:429–434

Jones SJ, Glen JW (1969) The effect of dissolved impurities on the mechanical properties of ice crystals. Philos Mag 19(157):13–24

Jouzel J (1984) Isotopes in cloud physics: multiphase and multistage condensation processes. In: Fritz P, Fontes JC (eds) Handbook of environmental isotope geochemistry, Vol 2. The terrestrial environment, B. Elsevier, New York, pp 61–112

Jouzel J, Merlivat L (1984) Deuterium and oxygen 18 in precipitation: modeling of the isotopic effects during snow formation. J Geophys Res 89(D7):11749–11757

Jouzel J, Souchez RA (1982) Melting-refreezing at the glacier sole and the isotopic composition of the ice. J Glaciol 28(98):35–42

Jouzel J, Merlivat L, Lorius C (1982) Deuterium excess in an East Antarctic ice core suggests higher relative humidity at the oceanic surface during the last glacial maximum. Nature 299(5885):688–691

Jouzel J, Lorius C, Petit JR, Jenton C, Barkov NI, Kotlyakov VN, Petrov VM (1987) A continuous isotope temperature record over the last climatic cycle (160000 years). Nature 329(6138): 403–408

Kamb B (1970) Sliding motion of glaciers: theory and observation. Rev Geophys Space Phys 8:673–728

Kamb B (1987) Glacier surge mechanism based on linked cavity configuration of the basal water conduit system. J Geophys Res 92(B9):9083–9100

Kamb B, LaChapelle E (1964) Direct observation of the mechanism of glacier sliding over bedrock. J Glaciol 5(38):159–172

Kaplyanskaya FA, Tarnogradsky VD (1976) Relict glacier ice in the north of Western Siberia and its role in the structure of regions of the Pleistocene glaciations of the cryolithozone, vol 231. Far-Eastern Section of the Academy of Sciences of the USSR, Vladivostok, p 5 (in Russian)

Kemper WD, Sills YK, Aylmore LA (1970) Separation of adsorbed cation species as water flows through clays. Soil Sci Soc Am Proc 34:946–948

Kittleman LR (1964) Application of Rosin's distribution in size-frequency analysis of clastic rocks. J Sediment Petrol 34:483–502

Koerner RM (1961) Glaciological observations in Trinity Peninsula, Graham Land, Antarctica. J Glaciol 3(30):1063–1074

Koerner RM (1970) Some observations on superimposition of ice on the Devon Island ice cap, NWT, Canada. Geogr Ann 52A:57–67

Koerner RM (1979) Accumulation, ablation and oxygen isotope variations on the Queen Elisabeth Islands ice Caps, Canada. J Glaciol 86:25–41

Koerner RM (1989) Ice core evidence for extensive melting of the Greenland Ice Sheet in the Last Interglacial. Science 244:964–968

Koerner RM, Fisher DA (1979) Discontinuous flow, ice texture, and dirt content in the basal layers of the Devon Island Ice Cap. J Glaciol 23(89):209–221

Kristensen M (1983) Iceberg calving and deterioration in Antarctica. Progr Phys Geogr 7(3):313–328

Lange MA, Ackley SF, Wadhams P (1989) Development of sea ice in the Weddell Sea. Ann Glaciol 12:92–96

Langway CC (1970) Stratigraphic analysis of a deep ice core from Greenland. Geol Soc Am Special paper 125:186

Lawson DE (1979) Sedimentological analysis of the western terminus region of the Matanuska Glacier, Alaska. CRREL Report 79-9, 122 p

Lawson DE, Kulla JB (1978) An oxygen isotope investigation of the origin of the basal zone of the Matanuska Glacier, Alaska. J Geol 86(6):673–685

Lemmens M, Lorrain R, Haren J (1982) Isotopic composition of ice and subglacially precipitated calcite in an alpine area. Z Gletscherkd Glazialgeol 18(2):151–159

Lliboutry L (1968) General theory of subglacial cavitation and sliding of temperate glaciers. J Glaciol 7(49):21–58

Lliboutry L (1987) Realistic, yet simple bottom boundary conditions for glaciers and ice sheets. J Geophys Res 92(B9):9101–9109

Lorius C (1968) A physical and chemical study of the coastal ice sampled from a core drilling in Antarctica. Commission of snow and ice, general assembly of Bern, September/October 1967. IAHS Publication 79, pp 141–148

Lorius C, Merlivat L (1977) Distribution of mean surface stable isotope values in East Antarctica: observed changes with depth in the coastal area. Isotopes and impurities in snow and ice. Proceedings of the Grenoble Symposium August/September 1975, I.A.H.S., Publication 118, pp 127–137

Lorius C, Vallon M (1967) Etude structurographique d'un glacier antarctique. CR Acad Sci Sér D 265:315–318

Lorius C, Raynaud D, Petit JR, Jouzel J, Merlivat L (1984) Late-glacial maximum-Holocene atmospheric and ice-thickness changes from Antarctic ice-core studies. Ann Glaciol 5:88–94

Lorrain R, Demeur P (1985) Isotopic evidence for relic Pleistocene glacier ice on Victoria Island, Canadian Arctic Archipelago. Arct Alp Res 17(1):89–98

Lorrain R, Souchez R, Tison J-L (1981) Characteristics of basal ice from two outlet glaciers in the Canadian Arctic – implications for glacier erosion. Curr Res Geol Surv Can Pap 81-1B:137–144

Mackay JR (1971) The origin of massive icy beds in permafrost, Western Arctic Coast, Canada. Can J Earth Sci 8:397–422

Mackay JR, Black RF (1973) Origin, composition and structure of perennially frozen ground and ground ice: a review. In: Permafrost. The North American contribution to the 2nd International Conference. National Academy of Science, Washington DC, pp 185–192

Macpherson D, Krouse HR (1967) O^{18}/O^{16} ratios in snow and ice of the Hubbard and Kaskawulsh glaciers. In: Stout GE (ed) Proceedings of the symposium on isotopes techniques in the hydrologic cycle. American Geophysical Union, Washington DC, pp 180–194

Malo BA, Baker RA (1968) Cationic concentration by freezing. In: Gould RF (ed) Trace inorganics in water. American Chemical Society. Advances in Chemistry Series 73, Washington DC, pp 149–163

McIntyre NF (1985) The dynamics of icesheet outlets. J Glaciol 31(108):99–107

Meier MF (1960) Mode of flow of Saskatchewan Glacier, Alberta, Canada. US Geol Surv Prof Pap 351:70

Meier MF, Post AS (1962) Recent variations in mass net budgets of glaciers in Western North America. Variations of the regime of existing glaciers. Proceedings of Obergurgl Symposium, September 1962. IAHS Publication 58, pp 63–77

Meier MF, Tangborn WV, Mayo LR, Post A (1971) Combined ice and water balances of Gulkana and Wolverine glaciers, Alaska, and South Cascade glacier, Washington, 1965 and 1966 hydrologic years. US Geol Surv Prof Pap 715A:23

Mellor M (1964) Snow and ice at the earth's surface. CRREL Monograph II-C1:163

Mellor M, Smith JH (1967) Creep of snow and ice. In: Oura H (ed) Physics of snow and ice. Proceedings of the Sapporo International Conference on low temperature science 1966, vol 1. Institute of Low Temperature Science, Sapporo, pp 843–855

Mercer JH (1978) West Antarctic ice sheet and CO_2 greenhouse effect: a threat of disaster. Nature 271:321–325

Merlivat L, Coantic M (1975) Study of mass transfer at the air-water interface by an isotopic method. J Geophys Res 80:3455–3464

Merlivat L, Ravoire J, Vergnaud JP, Lorius C (1973) Tritium and deuterium content of the snow in Greenland. Earth Planet Sci Lett 19:235–240

Michel FA, Fritz P (1982) Significance of isotope variations in permafrost water at Illisarvik, NWT. In: French HM (ed) Proceedings of the Fourth Canadian Permafrost Conference, Calgary. National Research Council of Canada, Ottawa, pp 173–181

Morgan VI (1972) Oxygen isotope evidence for bottom freezing on the Amery Ice Shelf. Nature 238:393–394

Moser H, Stichler W (1970) Deuterium measurements on snow samples from the Alps. In: Isotope hydrology. Proceedings of the Vienna Symposium 1970. International Atomic Energy Agency, Vienna, pp 43–57

Moser H, Stichler W (1975) Deuterium and oxygen-18 contents as index of the properties of snow blankets. Snow mechanics. Proceedings of the Grindelwald Symposium, April 1974. IAHS Publication 114, pp 122–135

Moser H, Stichler W (1980) Environmental isotopes in ice and snow. In: Fritz P, Fontes JC (eds) Handbook of environmental isotope geochemistry, Vol 1. The terrestrial environment, A. Elsevier, Amsterdam, pp 141–178

Müller F (1962) Zonation in the accumulation area of the glaciers of Axel Heiberg Island, N.W.T., Canada. J Glaciol 4:302–313

Mulvaney R, Wolff EW, Oates K (1988) Sulphuric acid at grain boundaries in Antarctic ice. Nature 331(6153):247–249

Nougier J, Lorius C (1969) Etude géologique et physico-chimique de carottes profondes de glace (Terre Adélie). Rev Géogr Phys Géol Dynam 11(2):165–170

Nye JF (1951) The flow of glaciers and ice sheets as a problem in plasticity. Proc R Soc Lond Ser A 207:554–572

Nye JF (1952) The mechanics of glacier flow. J Glaciol 2:82–93

Nye JF, Frank FC (1973) Hydrology of the intergranular veins in a temperate glacier. Proceedings of the Symposium on the hydrology of glaciers. Cambridge, September 1969. IAHS Publication 95, pp 157–161

O'Neil JR (1968) Hydrogen and oxygen isotope fractionation between ice and water. J Phys Chem 72(10):3683–3684

Østrem G (1965) Problems of dating ice-cored moraines. Geogr Ann 47A(1):1–38

Paren JG, Potter JR (1984) Isotopic tracers in polar seas and glacier ice. J Geophys Res 89 (C1):749–750

Paterson WSB (1977) Secondary and Tertiary creep of glacier ice as measured by borehole closure rates. Rev Geophys Space Phys 15(1):47–55

Paterson WSB (1981) The physics of glaciers, 2nd end. Pergamon, Oxford, 380 p

Peel DA, Mulvaney R, Davison BM (1988) Stable isotope/air temperature relationships in ice cores from Dolleman Island and the Palmer Land Plateau, Antarctic Peninsula. Ann Glaciol 10:130–136

Petit JR, Briat M, Rayer A (1981) Ice age aerosol content from East Antarctic ice core samples and past wind strength. Nature 293:391–394

Petit JR, Jouzel J, Pourchet M, Merlivat L (1982) A detailed study of snow accumulation and stable isotope content in the Dome C (Antarctica). J Geophys Res 87(C6):4301–4308

Philberth K, Federer B (1971) On the temperature profile and the age profile in the central part of cold ice sheets. J Glaciol 10(58):3–14

Picciotto E (1967) Geochemical investigations of snow and firn samples from East Antarctica. Antarct J US 2(6):236–240

Picciotto E, Maere X de, Friedman I (1960) Isotopic composition and temperature of formation of Antarctic snows. Nature 187:857–859

Picciotto E, Deutsch S, Aldaz L (1966) The summer 1957–1958 at the South Pole: an example of an unsusual meteorological event recorded by the oxygen isotope ratios in the firn. Earth Planet Sci Lett 1:202–204

Picciotto E, Crozaz G, Ambach W, Eisner H (1967) Lead-210 and strontium-90 in an alpine glacier. Earth Planet Sci Lett 3:237–242

Posey JC, Smith HA (1957) The equilibrium distribution of light and heavy waters in a freezing mixture. J Am Chem Ass 79(1):555–557

Potter JR, Paren JG, Laynes J (1984) Glaciological and oceanographical calculations of the mass balance and oxygen isotope ratio of a melting ice shelf. J Glaciol 30(105):161–170

Prantl F, Ambach W, Eisner H (1973) Alpine glacier studies with nuclear methods. The role of snow and ice in hydrology. Proceedings of the Banff Symposium, September 1972, vol 1. IAHS Publication 107, pp 435–444

Ragle RH, Blair RG, Person LE (1964) Ice cores studies on Ward Hunt ice shelf 1960. J Glaciol 5(37):39–59

Raymond CF (1971) Flow in a transverse section of Athabasca Glacier, Alberta, Canada. J Glaciol 10(58):55–84

Raymond CF, Harrison WD (1975) Some observations on the behaviour of the liquid and gas phases in temperate glacier ice. J Glaciol 14:213–233

Raynaud D (1976) Les inclusions gazeuses dans la glace de glacier; leur utilisation comme indicateur du site de formation de la glace polaire; applications climatiques et rhéologiques. Thèse de doctorat d'Etat. Université scientifique et médicale de Grenoble, 110 p

Raynaud D, Lebel B (1979) Total gas content and surface elevation of polar ice sheets. Nature 281(5729):289–291

Raynaud D, Whillans IM (1979) Total gas content of ice and past changes of the northwest Greenland ice sheet. Sea level, ice and climatic change. Proceedings of the Canberra Symposium IAHS Publication 131, pp 235–237

Raynaud D, Whillans IM (1982) Air content of the Byrd core and past changes in the West Antarctic ice sheet. Ann Glaciol 3:269–273

Redfield AC, Friedman I (1969) The effect of meteoric water, melt water and brine on the composition of Polar Sea water and of the deep waters of the ocean. Deep-Sea Res 16:197–214

Reeh N (1968) On the calving of ice from floating glaciers and ice shelves. J Glaciol 7(50):215–232

Reeh N, Thomsen H (1986) Stable isotope studies on the Greenland ice-sheet margin. Report of Activities 1985. Grønl Geol Under Rapp 130:108–114

Reheis MJ (1975) Source, transportation and deposition of debris on Arapaho Glacier, Front Range, Colorado, USA. J Glaciol 14(72):407–420

Ricq de Bouard M (1977) Migration of insoluble and soluble impurities in temperate ice: study of a vertical ice profile through the Glacier du Mont de Lans, French Alps. J Glaciol 18:231–238

Rigsby GP (1955) Study of ice fabrics, Thule area, Greenland. SIPRE Report 26:6

Robin G de Q (1955) Ice movement and temperature distribution in glaciers and ice sheets. J Glaciol 2:523–532

Robin G de Q (1974) Depth of water filled crevasses that are closely spaced (correspondence). J Glaciol 13:543

Robin G de Q (1976) Is the basal ice of a temperate glacier at the pressure melting point? J Glaciol 16(74):183–196

Robin G de Q (1977) Ice cores and climatic change. Philos Trans R Soc Lond 280:143–168

Robin G de Q (1979) Formation, flow and desintegration of ice shelves. J Glaciol 24:259–271

Robin G de Q (1983 a) Ice sheets: isotopes and temperatures. In: Robin G de Q (ed) The climatic record in polar ice sheets. Cambridge University Press, Cambridge, pp 1–18

Robin G de Q (1983 b) Profile data, Greenland region. In: Robin G de Q (ed) The climatic record in polar ice sheets. Cambridge University Press, Cambridge, pp 98–111

Rogers RR (1979) A short course in cloud physics. Inter Ser Nat Philos 96, Pergamon, Oxford, 235 p

Rosin P, Rammler E (1934) Die Kornzusammensetzung des Mahlgutes im Lichte der Wahrscheinlichkeitslehre. Kolloid Z 67:16–26

Röthlisberger H, Iken A (1981) Plucking as an effect of water-pressure variations at the glacier bed. Ann Glaciol 2:57–62

Röthlisberger H, Lang H (1987) Glacial hydrology. In: Gurnell AM, Clark MJ (eds) Glaciofluvial sediment transfer – an Alpine perspective. Wiley, Chichester, pp 207–284

Schwander J, Stauffer B (1984) Age difference between polar ice and the air trapped in its bubbles. Nature 311:45–47

Schwerdtfeger W (1970) The climate of the Antarctic. In: Orvig S (ed) Climates of the polar regions. World Survey of Climatology, vol 14. Elsevier, Amsterdam, pp 253–355

Sharp M (1985) Sedimentation and stratigraphy at Eyjabakka jökull, an Icelandic surging glacier. Quaternary Res 24(3):268–284

Sharp M, Gomez B (1986) Processes of debris comminution in the glacial environment and implications for quartz sand-grain micromorphology. Sediment Geol 46:33–47

Sharp RP, Epstein S, Vidziunas I (1960) Oxygen-isotope ratios in Blue Glacier, Olympic Mountains, Washington. J Geophys Res 1966 65(12):4043–4059

Shreve RL (1972) Movement of water in glaciers. J Glaciol 11:205–214

Shreve RL (1984) Glacier sliding at subfreezing temperature. J Glaciol 30(106):341–347

Smith VG, Tiller WA, Rutter JW (1955) A mathematical analysis of solute redistribution during solidification. Can J Phys 33:723–744

Solomatin VI (1981) On conditions of buried ice conservation in the permafrost zone. National Resources of West Siberia, vol 8. Moscow University Press, Moscow, pp 75–94 (in Russian)

Souchez R (1971) Ice-cored moraines in South Western Ellesmere Island, NWT, Canada. J Glaciol 10(59):245–254

Souchez RA, De Groote JM (1985) δD-$\delta^{18}O$ relationships in ice formed by subglacial freezing: paleoclimatic implications. J Glaciol 31(109):229–232

Souchez RA, Jouzel J (1984) On the isotopic composition in δD and $\delta^{18}O$ of water and ice during freezing. J Glaciol 30(106):369–372

Souchez R, Lemmens M (1985) Subglacial carbonate deposition: an isotopic study of a present-day case. Palaeogeogr Palaeoclimatol Palaeoecol 51:357–364

Souchez R, Lorrain R (1975) Chemical sorting effect at the base of an alpine glacier. J Glaciol 14(71):261–265

Souchez RA, Lorrain RD (1978) Origin of the basal ice layer from alpine glaciers indicated by its chemistry. J Glaciol 20(83):319–328

Souchez R, Tison J-L (1981) Basal freezing of squeezed water: its influence on glacier erosion. Ann Glaciol 2:63–66

Souchez R, Lorrain R, Lemmens M (1973) Refreezing of interstitial water in a subglacial cavity of an alpine glacier as indicated by the chemical composition of ice. J Glaciol 12(66):453–459

Souchez R, Lemmens M, Lorrain R, Tison J-L (1978) Pressure-melting within a glacier indicated by the chemistry of regelation ice. Nature 273(5662):454–456

Souchez R, Tison J-L, Jouzel J (1987) Freezing rate determination by the isotopic composition of the ice. Geophys Res Lett 14(6):599–602

Souchez R, Tison J-L, Jouzel J (1988a) Deuterium concentration and growth rate of Antarctic first-year sea ice. Geophys Res Lett 15(12):1385–1388

Souchez R, Lorrain R, Tison J-L, Jouzel J (1988b) Co-isotopic signature of two mechanisms of basal ice formation in arctic outlet glaciers. Ann Glaciol 10:163–166

Souchez R, Lemmens M, Lorrain R, Tison J-L, Jouzel J, Sugden D (1990) Influence of hydroxyl-bearing minerals on the isotopic composition of ice from the basal zone of an ice sheet. Nature 345:244–246

Stauffer B, Hofer H, Oeschger H, Schwander J, Siegenthaler U (1984) Atmospheric CO_2 concentration during the last glaciation. Ann Glaciol 5:160–164

Stauffer B, Fischer G, Neftel A, Oeschger H (1985a) Increase of atmospheric methane recorded in antarctic ice. Science 229:1386–1388

Stauffer B, Neftel A, Oeschger H, Schwander J (1985b) CO_2 concentration in air extracted from Greenland ice samples. In: Langway C et al (eds) Greenland ice core: geophysics, geochemistry and the environment. Geophysical Monograph 33, Am Geophys Union, Washington DC, pp 85–89

Stewart M (1975) Stable isotope fractionation due to evaporation and isotopic exchange of falling water drops: application to atmospheric processes and evaporation of lakes. J Geophys Res 80:1138–1146

Sugden D (1977) Reconstruction of the morphology, dynamics and thermal characteristics of the Laurentide Ice Sheet at its maximum. Arct Alp Res 9(1):21–47

Sugden DE, John BS (1976) Glaciers and landscape. Arnold, London, 376 p

Sugden DE, Knight PG, Livesey N, Lorrain RD, Souchez RA, Tison J-L, Jouzel J (1987) Evidence of two zones of debris entrainment beneath the Greenland Ice Sheet. Nature 328(6127):238–241

Swithinbank CWM, Darby DG, Wohlschlag DE (1961) Faunal remains on an antarctic ice shelf. Science 133:764–766

Terwilliger KP, Dizio SF (1970) Salt rejection phenomena in the freezing of saline solutions. Chem Eng Sci 25:1331–1349

Thomas RH (1979) Ice shelves: a review. J Glaciol 24(90):273–286

Thomas RH (1985) Responses of the polar ice sheets to climatic warming. In: Glaciers, ice sheets and sea level: effect of a CO_2 induced climatic change. Workshop held in Seattle, September 1984. US Department of Energy, pp 301–316

Thompson EG, Sayles FM (1972) In situ creep analysis of room in frozen soil. J Soil Mech Found Div 98:899–916

Thompson LG, Mosley-Thompson E (1981) Microparticle concentration variations linked with climatic change: evidence from polar ice cores. Science 212:812–815

Tiller WA, Jackson KA, Rutter JW, Chalmers B (1953) The redistribution of solute atoms during the solidification of metals. Acta Metall 1:428–437

Tison J-L, Haren J (1989) Isotopic, chemical and crystallographic characteristics of first-year sea ice from Breid Bay (Princess Ragnhild Coast-Antarctica). Antarct Sci 1(3):261–268

Tison J-L, Lorrain RD (1987) A mechanism of basal ice layer formation involving major ice-fabric changes. J Glaciol 33(113):47–50

Tison J-L, Souchez R, Lorrain R (1989) On the incorporation of unconsolidated sediments in basal ice: present-day examples. Z Geomorphol N.F. Suppl 72:173–183

Tsytovich NA (1957) The fundamentals of frozen ground mechanics. In: Proceedings of the 4th International Conference of Soil Mechanics and Foundation Engineering. London, 1957, vol 1, pp 116–119

Wagenbach D (1989) Environmental records in Alpine glaciers. In: Oeschger H, Langway C (eds) The environmental record in glaciers and ice sheets. Physical, chemical and earth sciences. Research Report 8. Wiley, New York, pp 69–83

Walder JS (1982) Stability of sheet flow of water beneath temperate glaciers and implications for glacier surging. J Glaciol 28:273–293

Walder JS (1986) Hydraulics of subglacial cavities. J Glaciol 23(89):335–346

Weeks WF, Ackley S (1986) The growth, structure and properties of sea ice. In: Untersteiner N (ed) The geophysics of sea ice, Nato ASI Series, Series B, Physics, 146. Plenum, Oxford, pp9–164

Weeks WF, Gow AJ (1978) Preferred crystal orientations along the margin of the Arctic Ocean. J Geophys Res 84(C10):5105–5121

Weertman J (1961) Mechanism for the formation of inner moraines found near the edge of cold ice caps and ice sheets. J Glaciol 3(30):965–978

Weertman J (1964) The theory of glacier sliding. J Glaciol 5(39):287–303

Weertman J (1966) Effect of a basal water layer on the dimensions of ice sheets. J Glaciol 6(44):191–207

Weertman J (1968) Diffusion law for the dispersion of hard particles in an ice matrix that undergoes simple shear deformation. J Glaciol 7(50):161–165

Weertman J (1973) Can a water-filled crevasse reach the bottom of a glacier? Proceedings of the Symposium on the hydrology of glaciers, Cambridge, September 1969. IAHS Publication 95, pp 139–145

Weertman J (1986) Basal water and high-pressure basal ice. J Glaciol 32(112):455–463

Weertman J, Birchfield GE (1982) Subglacial water flow under ice stream and west antarctic ice-sheet stability. Ann Glaciol 3:316–320

Weertman J, Birchfield GE (1983) Stability of sheet water flow under a glacier. J Glaciol 29:374–382

Weiss RF, Bucher P, Oeschger H, Craig H (1972) Compositional variations of gases in temperate glaciers. Earth Planet Sci Lett 16:178–184

Whalley WB, Krinsley DM (1974) A scanning electron microscope study of surface textures of quartz grains from glacial environments. Sedimentology 21:87–105

Wilcox WR (1964) Incomplete liquid mixing in crystal growth from the melt. J App Phys 35(3):636–643

Yurtsever Y, Gat JR (1981) Atmospheric waters. In: Gat JR, Gonfiantini R (eds) Stable isotope hydrology, deuterium and oxygen-18 in the water cycle. International Atomic Energy Agency. Technical reports series No 210, Vienna, pp 103–142

Zotikov IA (1986) The thermophysics of glaciers. Reidel, Dordrecht, 275 p

Subject Index

Aavatsmarkbreen 103
ablation rate 3, 8, 89, 114, 131, 141, 166, 176, 178
– zone 3, 13, 14, 16, 19, 20, 27, 68, 86, 90, 92, 96, 101, 102, 114, 115
abrasion 70, 71, 73, 133, 135, 138, 139
accumulation rate 4, 6, 8, 40, 41, 67, 80, 81, 82, 87, 89, 114, 166
– zone 3, 14, 19, 88, 90, 93, 101, 114, 123, 140
activity index 13, 23
adhesive bond 122
advection 115
aerosol 66
Agassiz Ice Cap 84
age measurement 41
aggregate 16, 124, 125, 139
air flow mechanism 145
Aktineq Glacier 131, 132
Alaska 4, 120, 158
Aletsch Gletscher 75
algae 172
alignment of crystals 170
alpine glacier 16, 24, 26, 36, 37, 65, 71, 140–163
– – flow 97
– permafrost 121
altitudinal effect 32
Amery Ice Shelf 24, 168, 172, 174
angle of internal friction 123
Antarctic Ice Sheet 24, 81, 116, 188
– Peninsula 7, 24, 35, 185, 188
Antarctica 4, 5, 9, 14–18, 24
apparent fractionation 128
– – coefficient 51, 56
Ar (argon) 77, 125
Arctic Canada 20, 27, 92, 104, 106, 108, 164
argon *see Ar*
Athabaska Glacier 9
Austrian Alps 68, 98
Axel Heiberg Island 19

Baffin Island 20, 27, 106
Banks Island 112
Barnes Ice Cap 20, 22, 27, 84, 85, 92, 93, 96, 106, 112
basal cavity 144
– freezing 103, 104, 116, 144, 176, 183
– heat 6, 115
– – flux 167
– ice 11, 20, 27, 70, 71, 74, 84, 104, 112, 113, 116, 123–126, 128–133, 135, 139, 143–145, 147–151, 154–161, 163, 166, 172, 181, 185, 188
– – accretion 103
– – layer 6, 106, 116, 147, 148, 151–154, 158
– melting 7, 8, 102, 104, 115, 117, 168, 186
– meltwater 104, 126
– pressure 143
– shear 16
– – stress 9, 13, 116, 120
– sliding 8, 11, 12, 118, 139, 143, 147
– stress 12, 13
– till 70
– water 11, 119, 141
– zone 7, 22, 74, 86, 113, 114, 123, 125, 126, 128, 140, 158, 185, 186
Bavarian Alps 36
bed bump 141
– deformation 142, 147
– obstacle 12, 116, 155
– protuberance 11, 14, 102, 136, 144, 155
Beta radioactivity measurement 36
BIL *see basal ice layer*
BLT *see boundary layer thickness*
Blue Glacier 141, 143
blue ice 84, 92, 93, 106, 128
bluish-grey ice 93
bottom freezing 28, 126, 164–166, 168, 170, 172, 174, 176, 182, 183, 185, 187, 188
– melting 27, 28, 164–166, 183, 185
boudinage 84
boundary condition 11, 115, 116, 168, 187, 188
– layer 54–56, 180
– – thickness 54, 180, 181

Breid Bay 24, 180
Breidamerkurjökull 119, 123
brine drainage 177, 181
– entrapment 181
– inclusion 170
– layer spacing 170
– plume 172
– pocket 170
bubble 4, 17–19, 26, 65, 74, 75, 85, 93, 125, 132, 134, 139, 141, 147, 151, 152, 156
– stratification 125
bubble-free ice 19, 74, 110, 125
bubbly ice 18, 19, 27, 74, 75, 84, 92, 106, 110, 131, 148, 177
buoyancy 165, 172
buried glacier ice 108, 110, 112, 113
Bylot Island 131
Byrd glacier 9, 24
– (ice) core 81, 123, 126
– Station 4, 5, 18, 24, 40, 68, 81, 125, 127, 128

C.A.R.O.L.I.N.E. ice core 128
Ca (calcium) 65, 66, 149–153, 156–158
calcite deposit 155–157
calcium *see Ca*
calving 3, 7, 164, 165
Camp Century 16, 18, 25, 67, 68, 82, 123, 126, 127
–– (ice) core 42, 81, 87, 112, 123–127
Canadian Arctic 49, 112, 131, 174
– Rockies 9, 155
Cape Prudhomme 128
carbon dioxide 75, 155, 156, 187, 188
–– partial pressure 156
carbonate deposit *see calcite deposit*
Casey 81
catabatic wind 177
c-axis 16, 22, 148, 170, 171
Central Greenland 117, 136
CFC 187
chemical sorting effect 153
climatic change 93, 116, 187, 188
closed system (for isotopes) 46, 48, 51, 67, 68, 98, 101, 154, 159, 160
clotted ice 135, 136
CO_2 *see carbon dioxide*
co-isotopic analysis 132
cold glacier 22, 75, 121, 123
– ice 6, 26, 75, 145
– patch 143, 144
Colle Gnifetti 37
columnar zone 170
comminution 70
compressive flow 22
congelation ice 60, 148, 170, 177, 181
– sea ice 170
controlling obstacle 11, 143
cosmic dust 68, 69
Coulomb law 123
crack 64, 70, 140, 177
creep closure 142
– rate 8, 11, 83, 84, 121, 165
Crête 87, 90, 136
crevasse 19, 20, 27, 42, 64, 66, 68, 140, 144, 148, 154
crushing 70, 71, 133
cryostatic pressure 8, 74, 141, 164
crystal 4, 5, 14, 16–19, 26, 27, 32, 34, 75, 83, 125, 126, 132, 133, 143, 144, 147, 148, 170–172, 181
– nucleation 172
– size 14, 17, 18, 26, 45, 84, 85, 128, 148
– growth 17, 18, 50, 51

Dailey Island 176
Davos 38
debris band 106, 124, 125, 147, 152
– concentration 124–126
– content 27, 104, 112, 135, 147
– flow 102
– lamination 132
– transport 71, 101
– zone 125, 128, 185, 186
debris/ice ratio 74
debris-loaded ice 17, 74
debris-poor basal ice 132
debris-rich ice 84, 102, 104, 106, 110, 113, 132, 148, 185
deformation rate 84
δ scale 29
$\delta\, ^{13}C$ 158
dendrite tip 171
deuterium excess 33, 52, 53, 131, 161–163
Devon (Island) Ice Cap 18, 27, 84
diffusion coefficient 44, 45, 54, 56, 128, 171
– phenomenon 128
dirt layer 18
dispersed facies 135, 136, 138, 139
Disraeli Fiord 177, 178
Distribution coefficient 64, 155, 157, 177
Dome C 17, 24, 36, 68, 86
Donnan exclusion 153
double diffusion 171, 172
drainage system 8, 64, 75, 141
dry-snow zone 5
Dye 3 16, 25, 84, 123, 127

E.G.I.G. line 87, 88, 90
effective bed 117

– viscosity 119, 120
electrical conductivity 174
Ellesmere Island 104, 164, 177, 183
englacial debris 102–104
– temperature 167
equilibrium fractionation coefficient 29, 31, 33–35, 46, 47, 54, 128, 159, 172, 175, 181
– line 3, 4, 13, 23, 27, 88, 89, 96, 101, 102, 116, 166
eutectic composition 62
– temperature 61, 62, 64, 83, 166
evaporation effect 33
exchangeable ions 157

Fairbanks 120
fast ice 177
Fick's law of diffusion 180
Filchner-Ronne Ice Shelf 24, 26, 27
firn 4, 5, 26, 40–42, 44, 45, 66, 74, 75, 79,81, 88, 96, 98–101, 140, 187
– aquifer 140
– layer 67, 98
– line 140
firn-ice transition 4
Fletschhorn 145, 146
floating glacier 164, 165, 168
flow law 8, 11, 22, 121, 123
– line 21, 22, 24, 75, 79, 86, 90, 92, 93, 97, 101, 102, 112, 136, 141, 144, 166, 182
– model 124, 141
flushing out 65, 150
fold 22, 93, 123, 132, 148
folded ice 22
folding 22, 93, 102, 123, 128, 148
fractionation during melting 46, 138
frazil ice 170–172, 185
freezing experiment 58
– front 49, 54, 59, 108, 111, 157, 170, 179, 181
– interface 118, 179
– kinetics 128
– point 64, 117, 119, 152, 155, 166, 167, 171, 172, 182, 185, 188
– rate 26, 46, 50, 51, 54–56, 59, 60, 75, 126, 128, 138, 165, 169, 170, 174, 179–183, 188
– slope 46, 48, 49, 51, 52, 111, 129, 130, 132, 135, 138, 139, 159, 160, 163
– surface 119
– temperature 61, 155, 166
freezing-on 123–125, 158, 167, 170, 174, 175, 183, 185
French Alps 66, 152
frictional heating 64, 116
frozen-unfrozen sediment interface 123

gas analysis 125
– bubble 4, 75, 139, 156
– composition 75, 125, 141
– content 74, 75, 79, 81, 83, 92, 125, 128, 141
– diffusion 18, 125
– inclusion 124
– volume 79
George VI Ice Shelf 24, 185
geothermal heat 6, 7, 8, 114, 116, 141
girdle (ice fabrics) 15, 16
glacial permafrost 113
– transport 70
glacier bed 11, 15, 20, 70, 102, 103, 138, 140–143
Glacier d'Argentière 146, 152, 153
– de Tsanfleuron 146, 148, 153, 154, 156, 157, 159
– de Tsijiore Nouve 71, 146, 148, 150, 157
glacier ice 3, 4, 14, 18, 19, 26, 27, 45, 64, 65, 75, 92, 108, 111–113, 121, 122, 125, 126, 128, 130, 131, 135, 138, 147–151, 154, 158, 160, 161, 163, 174, 183, 184, 188
– sole 9, 70, 102, 113, 114, 116, 119, 123, 138, 142, 147, 148, 162, 188
– substrate interface 68
global change 187
grain boundary 60, 64, 65, 83, 84, 141, 144, 170
– growth 84
– rotation 16
– size distribution 70, 71, 110
granulo-viscous effect 122
greenhouse effect 187
– gas 187
Greenland Ice Sheet 10, 25, 26, 40, 68, 74, 81, 82, 84, 87, 90, 93, 127, 138, 139
Griesgletscher 141, 146
ground ice 108, 112
– moraine 145, 147, 162
grounding line 165, 166, 169, 174, 182, 183, 185, 186, 188
growth rate 60, 170, 181
Grubengletscher 49, 145–147, 160, 163

heat conduction 183
– flux 8, 114, 116, 144, 167, 185
– pump mechanism 143
– transfer model 58
hoarfrost 5, 35, 38
Holocene 18, 68, 69, 84, 86, 90, 93, 112
horizontal component of the velocity 103, 114
– transport of heat 115
hydroxyl-bearing mineral 73, 138, 139

ice accretion 103, 148, 152, 178, 187
– cap 3, 6, 16, 18, 20, 22, 24, 26, 27, 68, 79, 84–86, 92, 96, 104, 106, 114, 123, 131, 140, 147, 179, 183, 187
– chemistry 148, 157
– cliff 22, 26, 103, 131
– coating 148
– conduit 140
– core 14, 15, 18, 41, 49, 58, 65, 68, 81, 87, 92, 111, 123–125, 127, 138, 141, 170, 174, 177, 179–181, 188
– cored moraine 106, 108
– creep 83, 120
– dammed lake 145, 147, 162
– divide 18, 20, 87, 116, 123, 136
– dome 24
– doping 83
– fabrics 14, 16, 131, 148
– facies 132, 138, 186
– flow 3, 8, 9, 13, 19, 22, 68, 77, 79, 81, 82, 86, 96, 108, 132, 141, 155
– foliation 18, 20, 21
– laden sediment 121
– matrix 138
– nucleation 172
– perched moraine 105, 106, 108
– platelet 170
– ramp 131
– residence time 13, 14, 23, 68, 101
– rise 164, 188
– segregation 147
– sheet 3–10, 16, 18, 23, 24, 26, 27, 40, 60, 67, 68, 74, 79, 81, 82, 86–88, 96, 108, 113, 114, 116, 118–120, 124, 125, 130, 132, 140, 164, 165, 170, 174, 187, 188
– – profile 79, 81
– – sliding 113
– shelf 26, 164, 172, 174, 176–179, 188
– stream 12, 13, 24, 90, 119, 167, 188
Ice Stream B 13, 119
ice texture 147
– thickness 9, 18, 22, 81, 90, 115, 156, 165–167, 185
– type 22, 26
– warming 168
– water interface 54, 143, 167, 180
ice-bed interface 125, 128, 141, 143, 147, 148, 187
Iceland 42, 68, 104, 119
ice-wedge 110, 112
impurity 17, 28, 60, 61, 64–66, 83, 86, 188
– distribution 172
– leaching 64, 66
incorporation mechanism 123
initial transient 54, 56, 59, 138
– water 46, 54, 58, 128, 130, 132, 138, 158, 161–163
inland effect 32
interface position 56, 58
intergranular channel 141
– friction 122
– void 123
internal deformation 8, 9, 116
– flow 116
– melting 8, 75
isochron 101, 128
isotopic change 39, 45, 132, 159
– distribution 54, 172, 179, 180
– exchange 34, 40, 42, 44, 73, 139
– fractionation 29, 31, 33, 44, 45, 159
– homogenization 40, 42
– kinetic effect 34

Jakobshavn 87, 138, 139
– Isbrae 25, 90, 93, 96, 132, 135

K (potassium) 65, 149, 150–153
Kesselwandferner 37, 67, 98, 101
Koettlitz Ice Tongue 24, 168, 175

Lambert Glacier 174
Last Glacial Maximum 81, 82, 86, 90, 112, 113
last interglacial 128, 139
latent heat 5, 11, 64, 106, 113, 116, 143, 144, 166, 188
latitudinal effect 32
Laurentide Ice Sheet 82, 117
Law Dome 24, 81
leaching process 66
lead 210 66, 68, 98, 153
liquidus 166, 167
lithological composition 106
Little America 24, 37, 40, 168
log normal distribution 71

Mac Murdo *see McMurdo*
magnesium *see Mg*
marginal accretion 153
– area 130
– lake 133
– zone 14, 24, 90, 130, 132, 138, 139, 145
Marie Byrd Land 24, 68
marine ice transgression hypothesis 179
mass balance 3, 22, 101
Matanuska Glacier 158
Maudheim 24, 168
McMurdo 24, 168, 174
melt crust 26
melting experiment 65, 138, 151
– interface 118

– point isotherm 117, 119
– rate 142, 165, 169
Melville Bay 25, 82
Meserve Glacier 14
meteoric water line 33, 52, 132
methane 187
Mg (magnesium) 65, 66, 149, 150, 151, 153
microparticle 18, 68, 86
Milcent 88
milling experiment 70
mineral particle 65, 68, 70, 73, 101, 148, 183
molecular diffusivity 34
moraine 88, 90, 105, 106, 108, 112, 128, 145, 147, 162
moulin 64, 68, 140, 144
mud clot 125, 138, 139
multiple maxima (ice fabrics) 16, 17

N_2 (nitrogen) 75
Na (sodium) 60, 64–66, 92, 128, 149–153, 171, 180, 181
N-channel *see Nye channel*
net ablation 3, 8, 166
– balance gradient 13
Nuna Ramp 16
nunatak 7, 68, 70, 106
Nye-channel 141

O_2 (gas) 75
ocean water 28–30, 33, 166, 174
oceanic warming 188
open system (for isotopes) 48, 51, 159, 162
optic axis 14, 125
organic remains 174
Oscar II land 103
outlet glacier 10, 12, 24, 131, 132, 188

Pakitsup 93, 96
palaeoclimatic effect 32
– indicator 158
– reconstruction 113
– record 187, 188
Palmer land 185
partial freezing 148, 155, 158, 166
– melting 61, 154
^{210}Pb 66, 67, 97–101, 154
percolation 7, 26, 42, 45, 65, 97, 119, 154, 187
– flow 119
– zone 5, 7, 26
perfect gas law 79
permafrost 108, 113, 119–121, 145
– layer 118
phase change 7, 45, 143, 147, 166, 172
diagram 61
– equilibria 60
Plateau Station 24, 58
Pleistocene 18, 69, 74, 84, 85, 90, 93, 96, 106, 112, 113, 117, 128, 136
plucking 122
polynia 171
pore volume 79
– water pressure 119, 123
potassium *see K*
precipitation effect 32, 111
pressure gradient 64, 151–153, 156
– melting and regelation mechanism 143
– – point 6, 10, 11, 18, 102, 114, 117, 136, 138, 144
Prince Albert Peninsula 108
pure diffusion 56
– shear 20

Qigssertâq 25, 74, 138, 139

radar investigation 174
radioactive decay 101
– product 66
rate of deformation 116
Rayleigh distillation 30, 31
– model 32, 34, 128, 159
– process 46
– type distribution 54, 56, 59, 128
R-channel *see Röthlisberger-channel*
recrystallization 4, 16–18, 40, 44, 45, 61, 64, 65, 141
refreezing of meltwater 5, 44, 106, 126, 145
regelation ice 138, 144, 156, 157
– layer 143
– mechanism 136, 143, 144
– sliding 11, 141, 155, 157
rejection of solutes 56, 64
relic Pleistocene glacier ice 112
rheological contrast 85, 86
– properties 69, 117
rime ice 7
rock avalanche 102
– glacier 121, 145
Roi Baudouin Station 24, 35, 58, 67
Roosevelt Island 24, 174
Rosin's law of crushing 71
Ross Ice Shelf 24, 26, 168, 170, 174, 182
Röthlisberger-channel 141, 142
Russell Glacier 25, 132, 136, 138, 139

salinity 166, 167, 171, 174, 177–179, 181, 183–185
salt rejection 172
Sandhills Moraine 112
sandy loam permafrost 120
Saskatchewan glacier 9

Scandinavian Ice Sheet 82
sea ice shelf 176
– level 26, 42, 69, 83, 92, 128, 165, 187, 188
seasonal temperature 6, 79, 168
– variations 35, 41, 114
sediment strength 123
segregation ice 108, 110, 112, 113
selective electrolyte filtration 153
– flushing out of ions 150
self purification 64
semi permeable membrane 152
Seward glacier 4
shear deformation 84, 116, 138
– plane 11, 102, 122
– zone 102, 106
simple shear 16, 18, 20, 138
sliding interface 117
SMOW 29, 46, 158, 172, 174, 175, 178, 183
snow aquifer 140
snowflakes 68
sodium *see Na*
solidus 167
soluble impurity content 128
solute concentration 155
– redistribution 157
Søndre Srømfjord 25, 132
South Pole 36, 67
^{90}Sr (strontium 90) 98–100
stagnant ice mass 114
Station Centrale 87
steady state 13, 22, 54, 56, 89, 90, 114, 138, 142, 185
storm wave 165
strain rate 8, 9, 14, 16, 20–22, 87, 120, 123
stratified debris 126, 139
– facies 134, 135, 139, 158, 159, 186
streaming flow 132
stress 4, 8, 9, 11–14, 16, 19, 20, 22, 27, 116, 120, 123, 125, 143–145, 148, 164–166, 185
stress-strain relationship 120
strontium 90 *see* ^{90}*Sr*
subglacial aquifer 141
– cavity 148, 152, 153, 155
– channel 140, 145
– erosion 108, 116
– permafrost 145
– precipitate 155, 158
– sediments 8, 11, 117, 119, 152
– stream 145
– tunnel 151, 154, 160
– water 11, 13, 123, 129, 142, 144, 155, 156, 158, 162
supercooled water 183
supercooling 170
superimposed ice 3, 6, 7, 26, 27, 75, 93, 96, 106, 139
supraglacial stream 68, 150
surface melting 5–7, 113
surging 104, 142
Svalbard 103, 104
Swiss Alps 71, 75, 141, 142, 148, 153, 160
Sydkap Ice Cap 104, 106

Tanquary Fiord 183, 184
temperate ice 6, 65, 188
– glacier 22, 64, 75, 104, 123
temperature distribution 106, 114, 116, 168, 176
– gradient 7, 35, 40, 79, 114, 116, 155, 167
– inversion 35
– regime 103, 114
tensile force 164
– stress 165
tephra layer 68
terminal grade 70
– transient 54
Terre Adélie 15, 16, 24, 79, 81, 99
thermal conductivity 185
– diffusivity 171
– zone 116, 117
thermohaline convection 172
thickening rate 166
thin film 119, 122, 142, 153
thinning rate 166
Thule 16
tidewater glacier 164, 166, 185
till 113
– cohesion 123
– deformation 119
total gas content 75, 81, 83, 125, 128
tracer 73, 106
tractive force 70, 123
triple junction 60, 61, 64, 144
– point 61
tritium 29, 178
tubular bubble 26
tunnel 11, 14, 120, 121, 145, 147, 148, 151, 152, 154, 160

unconsolidate sediments 104, 117–119, 124, 125, 138
unidirectional freezing 49, 54, 56, 170
Unteraargletscher 142, 146

valley glacier 3, 10, 24, 27, 68, 102
Vatnajökull 42
vein flow 65
– system 64, 65, 141
– wall 64
vertical velocity component 86, 101, 114

-- vector 114, 115
Victoria Island 49, 108, 112
viscoplatic parameter 120
Vostok 4, 18, 24, 81, 189

Wallis 145
Ward Hunt Ice Shelf 164, 177-179
water-filled cavity 141, 142
- crevasse 19, 140
water film 138, 141-143, 156, 167
- flow 64, 140
- pressure 13, 104, 119, 123, 141, 142, 144, 147, 164
- seepage 141
- squeezing 150
Weddell Sea 24, 180
Weissfluhjoch 38
West Antarctica 18, 41, 68, 81, 119, 125
West Greenland 16, 87, 90, 123, 132
Western Canada 42, 109, 112
Western United States 42, 141, 143
wet-freezing zone 113
wet-snow zone 6
White Glacier 19, 22
white ice 84, 92, 93
wind blown snow 23
- drift 3, 37, 96

x-ray analysis 60, 68, 125

Zugspitzplatt 36